Ch. Vella   S.W. Ketteridge (Eds.)

Canine Parvovirus: A New Pathogen

Springer-Verlag Berlin Heidelberg GmbH

Dr. Cherelyn Vella
National Institute for Biological Standards and Control
Division of Virology
Blanche Lane, South Mimms
Potters Bar, Hertfordshire, EN 6 3 QG, UK

S. W. Ketteridge
School of Biological Sciences
Queen Mary and Westfield College
University of London
Mile End Road
London, E1 4NS, UK

ISBN 978-3-540-54314-5   ISBN 978-3-642-76797-5 (eBook)
DOI 10.1007/978-3-642-76797-5

10/3140/543210 – Printed on acid-free paper

Contents

1 Introduction

Canine parvovirus (CPV) is a new virus that was first reported early in 1978, responsible for scattered outbreaks of enteric disease in domestic dog populations of all ages. Within a few months this virus had crossed national and continental boundaries. It passed through quarantine barriers, causing many fatalities, at first in animals in breeding and boarding kennels, puppy farms and research facilities, later spreading to home-based dogs. By 1981, enteric disease caused by CPV had become world-wide, and a second manifestation, myocarditis, was recognised in young puppies only. At the time of its sudden appearance in 1978, canine populations simply had no immunity to CPV, and it consequently led to a disease of epizootic status with no prediliction for either sex or breed.

CPV is a member of the family of animal DNA viruses known as the Parvoviridae. The virus particles, which are amongst the smallest of the known viruses infecting vertebrates, are 23–26 nm in diameter (Paradiso et al. 1982). Like other parvoviruses, it is extremely resistant to inactivation in the environment, and the stability of shed virus particles is probably one of the most important factors in its transmission. CPV has not been formally classified by the International Committee for Virus Nomenclature, but according to Pollock and Parrish (1985) it is likely that it will be designated as CPV-2 to distinguish it from minute virus of canines (CPV-1), another parvovirus isolated earlier from dogs (Binn et al. 1970).

Other economically important parvovirus infections of livestock, domestic and sylvatic animals are known, including those of cattle (Abinati and Warfield 1961), mink (Schofield 1949), raccoons (Nettles et al. 1980), pigs (Johnson and Collings 1969), fowl (Parker et al. 1977) and cats (Cockburn 1947). In general, young animals appear to be much more susceptible; infection of adults is usually asymptomatic. However, CPV, feline panleucopenia virus (FPV) and mink enteritis virus (MEV) are distinguished by their ability to cause disease in all age groups. In recent years, evidence has also accumulated implicating parvoviruses in an expanding spectrum of human disease (reviewed in chapter 6).

The scale of the disease problem caused by CPV demanded immediate action and the outcome of the research stimulated has been impressive (for reviews see Berns 1984, Pollock and Parrish 1985, Parrish 1990). However, there are still many aspects of the biology of CPV that remain unclear, not least of which is the question of the possible origins of this virus.

1.1 Historical Background

Enteric parvo-like viruses were first isolated from dogs as early as 1967 (Binn et al. 1970). These were recovered from the stools of four clinically normal military dogs and serological analysis revealed specific neutralising antibodies in a high percentage of similar animals. Comparable antibody titres (1 : 4096) were also found in commercial canine gamma globulin preparations (Globulon, Pitman-Moore Co., Indianapolis, USA). The virus was detected by electron microscopy, the particles having a diameter of 20–21 nm, many with a hexagonal profile. Other physicochemical properties were also consistent with the parvovirus group. Siegl (1976) has since shown that this virus, named minute virus of canines (MVC, CPV-1), is widespread in healthy dogs and indeed the faecal shedding of other parvoviruses is now well documented (Bouillant and Hanson 1965; Csiza et al. 1971a). MVC has been antigenically and genetically distinguished from CPV (Carmichael et al. 1980; Macartney et al. 1988) and has been designated as CPV-1 by some authors (Pollock and Parrish 1985).

In 1977, another parvo-like virus was isolated from a litter of Borzoi puppies (Eugster and Nairn 1977). At birth these pups appeared normal, but they developed a non-haemorrhagic diarrhoea at 9 days and subsequently recovered, so that by day 20 they passed formed stools. Virus was purified from faecal samples and propagated in cultures of MDCK (Madin-Darby canine kidney) cells. A cytopathic effect with total cell lysis was seen on the first passage, but these effects were minimal on the second and lost on subsequent passage. In primary canine cells cytopathic effects were not observed. Electron microscope (EM) studies on the faecal samples and cell culture fluids showed hexagonal particles of 18–20 nm diameter.

The pathogenicity of these two parvo-like viruses is unknown. Unfortunately, Eugster and Nairn's isolate cannot be evaluated because the virus was subsequently lost in cell culture. However, Binn's isolate (MVC) could only be propagated in 1 of 22 cell lines and was not lethal for newborn or weaning mice, hamsters, guinea pigs or rabbits. Furthermore, the fact that this virus is widespread in healthy dogs, and that MVC-specific antibodies were present in canine blood products, suggests that MVC is normally non-pathogenic.

The first well documented cases of CPV were recorded between spring and autumn of 1978 in the United States of America (Appel et al. 1979), Canada (Thomson and Gagnon 1978) and Australia (Walker et al. 1979), and within a year in Europe. Afshar (1981) lists reports on the world-wide occurrence of CPV. Originally, the disease was recognised as a severe enteritis with vomiting. Morbidity rates approached 100% with mortality rates in the range 20%–80%. Young pups were particularly susceptible, but the disease was seen in dogs of all ages. A second syndrome, myocarditis, was found in pups that were infected *in utero* or neonatally, and death often occurred without premonitory symptoms.

Analysis of infected tissues and secretions confirmed that both disease syndromes were caused by CPV, which was assigned to the parvovirus group on the basis of its appearance in the EM (Else 1980), haemagglutination pattern (Appel et al. 1979) and resistance of virions to heat, pH 3.0 and ether (Johnson and Spradbrow 1979). Furthermore, CPV cross-reacted with FPV, a well-characterised parvovirus of cats.

Testing of canine sera stored for about 10 years prior to 1976 showed no trace of CPV-specific antibodies, but these were easily detected after that date. On this evidence, CPV was accepted as a genuinely new pathogen.

During the initial panzootic the effect on dogs was devastating. All breeds of domestic dog were susceptible, although recent reports have suggested that Doberman Pinschers, Rottweilers, Setters, Pointers and German Shepherds are more susceptible to CPV enteric disease (Glickman et al. 1985; Ernst et al. 1988; Rogers et al. 1987). Sylvatic canidae, including wolves (Fletcher et al. 1979), coyotes (Everman et al. 1980; Thomas et al. 1984), bush dogs and crab-eating fox (Mann et al. 1980) are also infected. In the United Kingdom, epizootics were first seen in the winter of 1978–1979 among dogs in kennels, puppy farms and research facilities where the confined conditions facilitated transmission (Hitchcock and Scarnell 1979). By the summer of 1980, CPV had spread to the home-based pet dog, resulting in the epizootic of that year (McCandlish et al. 1980). By 1981, the disease was enzootic and CPV was commonly seen in the small animal practice. During the years 1979 to 1982 the original virus was replaced worldwide by an antigenically and genetically variant virus designated CPV-2a. The epidemiological advantage of the new isolate over CPV has yet to be determined since there are no apparent differences in transmissibility and pathogenicity in both domestic and sylvatic canidae (Parrish et al. 1985, 1988a).

Infections caused by FPV have been known for about 100 years (Herringham and Andrews 1888; Zschokke 1900). This virus is both highly infectious and virulent causing high mortality in unvaccinated domestic cats. However, the disease is more severe in kittens (Fastier

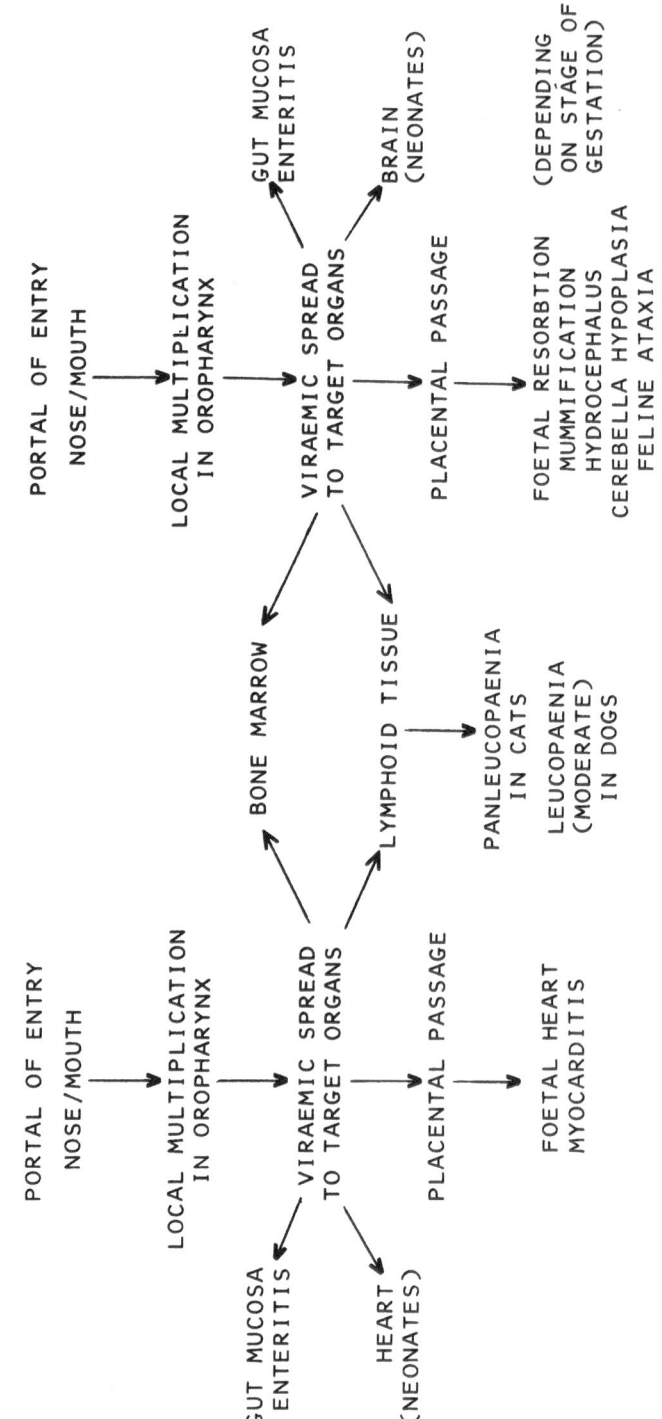

Fig. 1. Comparison of the pathogenesis of canine parvovirus (CPV) and feline panleucopenia virus (FPV)

1968). Lethal FPV infections have also been reported for tigers, leopards, cheetahs, lynxes, servals and ocelots at the London Zoo (Cockburn 1947). The structure and properties of virions of FPV and CPV have been found to be very similar. Furthermore the pathogenesis of the two virus infections is remarkably similar (Fig. 1), although *in utero* infection of kittens by FPV leads to feline (juvenile) ataxia, whereas with dogs CPV causes myocarditis (Csiza et al. 1971b).

Raccoon parvovirus (RPV) was found in 1938 (as reported by Waller 1940) in wild raccoons living near cat farms where FPV was enzootic. By 1940, an enteric disease similar to that caused by FPV was seen on fur farms, and Waller showed that the disease could be prevented in exposed raccoons with anti-FPV serum. The disease caused by RPV is now enzootic in North American raccoons (Nettles et al. 1980).

In 1947, a disease in minks caused by MEV was reported from mink farms all over Canada, and quickly spread to other parts of the world (Schofield 1949). Other captive mustelidae, including the otter, marten and badger, have been infected with MEV, which is considered to be a variant of FPV. In fact, early evidence suggested that RPV and MEV were antigenically indistinguishable from FPV (Johnson 1967; Waller 1940; Scott et al. 1970).

The early use of FPV vaccines in dogs, before CPV vaccines were available, did afford some protection against CPV. In 1981, Duphar produced an homologous inactivated CPV vaccine under the name "Kavac Parvo" and by 1984 modified live (attenuated) CPV vaccines were finally achieved. A change in the pattern of the disease has occurred over the years since 1980, with the development of immunity in canine populations, such that nowadays the majority of clinical cases present with the enteric syndrome.

1.2 Virological Considerations

CPV is a species of the genus *Parvovirus* (Siegl 1976), which in turn is a member of the family Parvoviridae (Bachmann et al. 1979). CPV belongs to *Parvovirus* sub-genus A, the group of autonomous parvoviruses that is distinguished by its ability to replicate in actively dividing cells without co-infection with an unrelated helper virus. Current ideas on the replication of parvovirus DNA have been reviewed by Hauswirth (1984) and Berns et al. (1985).

The virions of CPV (Fig. 2) are non-enveloped, icosahedral particles with diameters ranging from 23 to 26 nm in negatively stained

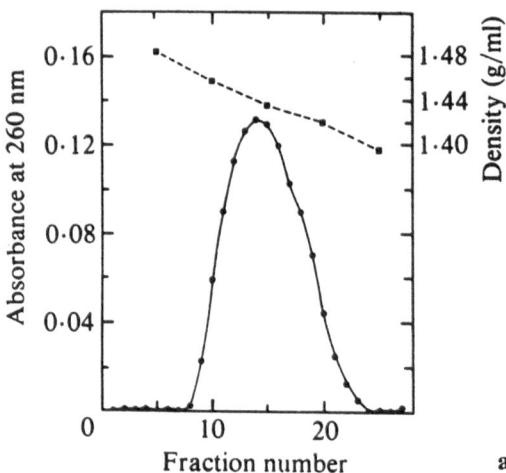

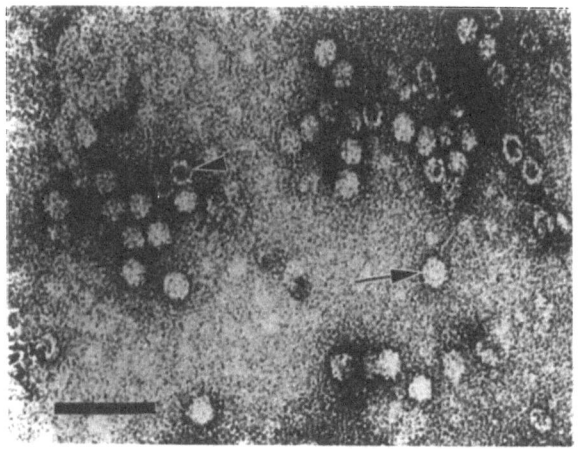

Fig. 2a, b. Virions of canine parvovirus (CPV). *a* Buoyant density analysis. Purified CPV was mixed with 20 m*M* TRIS pH 8, 1 m*M* EDTA, 0.2 % Sarkosyl and 1.44 g/ml CsCl and centrifuged at 110000 g for 48 h at 10 °C. Fractions (——) were collected from the bottom of the tube and the densities (– – –) determined by refractometry. *b* Electron micrograph of CPV virions negatively stained with uranyl acetate. *Arrow* indicates a particle that excluded the stain, and the *arrowhead* shows a capsid penetrated by uranyl acetate. *Bar marker* represents 100 nm (Paradiso et al. 1982)

preparations in the EM. Complete virions band in CsCl gradients with a buoyant density of 1.44 g/ml (Paradiso et al. 1982). Virus-specific enzymes are not present in the virions. Particles of CPV characteristically cause haemagglutination of red blood cells from a number of animals, including pig, rhesus monkey, African green monkey, cat and crab-eating macaque (Carmichael et al. 1980).

The viral genome is composed of linear single-stranded DNA and is of minus polarity with respect to the transcription of viral mRNA. Like other autonomous parvoviruses, CPV differs from the helper-dependent parvoviruses (e.g. adeno-associated virus AAV) in that it encapsidates >99 % strands of only one polarity. In contrast, AAV encapsidates strands of both polarities with equal frequency into separate virions (Mayor et al. 1979; Berns and Adler 1972). The virion DNA is composed of 4900 ± 100 nucleotides (McMaster et al. 1981; Tratschin et al. 1982; Reed et al. 1988) and the ends of the molecule contain palindromic sequences which are though to be involved in priming replication of the viral genome leading to the production of replicative form (RF) DNA. The organisation of the genome of autonomous parvoviruses is relatively simple, and there are probably just two separate genes: one codes for virion structural proteins and the other for a non-capsid protein of uncertain function that can be detected in infected cells and is possibly involved in the replication of viral DNA (Pintel et al. 1983; Rhode and Paradiso 1983).

The capsid of CPV particles is typical of the parvoviruses and is constructed from three structural proteins of molecular weight 82 300 (VP1), 67 300 (VP2') and 63 500 (VP2) (Paradiso et al. 1982). One of the characteristics of the capsid is that protein VP2 is formed from VP2' by proteolytic cleavage, and this normally occurs after assembly of the virions. VP2 is the major capsid protein. If CPV is analogous with FPV and minute virus of mice (MVM), then the amino acid sequences of the three structural proteins form a nested set. The entire sequence of VP2 is contained within VP2', which in turn is contained within the VP1 protein (Tattersall et al 1977; Carlson et al. 1985).

Virions of CPV are further characterised by their stability and resistance to disruptive treatments. CPV is relatively resistant to heat inactivation; there is little loss of infectivity after exposure to 60 °C for 1 h. However, CPV may not be so heat resistant as FPV, since it is inactivated by heating to 80 °C for 1 h, whereas FPV requires 2 h under similar conditions (Goto et al. 1984). Infectivity of CPV is not destroyed by common disinfectants, including acids, bases, ether, chloroform, phenols and quarternary ammonium compounds. However, 0.2 % formalin and sodium hypochlorite bleach (1 : 30 parts in water) are effective disinfectants (Kramer et al. 1980). It has been previously claimed that CPV retains its infectivity in faeces at room temperature for 6 months (Pollock and Carmichael 1981). However, a more recent report suggests that infectivity is reduced considerably after 2 months' storage at room temperature. No apparent reduction in infectivity occurs over 12 months' storage at a temperature of −20 °C (Gordon and Angrick 1986).

In infected cells CPV characteristically multiplies in nuclei and the cytopathological changes (margination of the heterochromatin, vacuolation of the nucleolar fibrous components, and association of virus capsids with chromatin fibres) are considered to be diagnostic of parvovirus infection (Paradiso et al. 1982).

2 Canine Parvovirus Disease Syndromes

There are two distinct disease syndromes caused by CPV infection. These are the enteric form, a haemorrhagic enteritis, and the cardiac form, generally described as a non-suppurative myocarditis. A third manifestation, generalised necrotizing vasculitis, is occasionally diagnosed in puppies (Johnson and Castro 1984). With the pig parvovirus (PPV) and some other parvoviruses, infection of breeding stocks can lead to a further syndrome resulting in reduced fertility. However, despite worries voiced by owners and breeders (e.g. Thompson et al. 1985), there is no evidence that either CPV or its vaccine strains seem to be linked with these problems in dogs. The pathology of the disease is related to the age at which the animal succumbs to infection. Sites of virus multiplication are determined to a large extent by the fact that virus replication can only be supported by host cells which are actively dividing and so may be modified by any factor which influences the mitotic index of host tissue.

2.1 Canine Parvovirus-Induced Myocarditis

In utero or neonatal infection results in the cardiac form of the disease. This has been correlated with the rapid expansion of the myocardium during the later stages of foetal development, and with the fact that during the first 9 days of life, 2 %–4 % of myocytes are undergoing mitosis (Bishop and Hine 1975). There are three clinical presentations:

1. Acute non-suppurative myocarditis is the most common clinical form, seen in very young pups of up to 8 weeks of age. Death occurs suddenly, frequently without premonitory symptoms, and is thought to be due to terminal cardiac arrhythmias (Kelly and Atwell 1979; Atwell and Kelly 1980).

2. Subacute heart failure with respiratory distress is seen in pups over the age of 8 weeks. In general, there is a sudden onset of dyspnoea followed by collapse. Symptoms which include tachycardia and cardiomegaly may be demonstrated on X-rays. Death usually follows within 24–48 h due to severe multifocal necrosis (Kelly and Atwell 1979; Robinson et al. 1980).
3. Cardiac myopathies and congestive heart failure are often seen in adolescent pups and adult dogs of between 5 months and 3 years. Clinical symptoms follow a history of exercise intolerance and periodic anorexia and include irregular pulse, regurgitant heart murmur, and moist crepitant rale of the lungs. There is abdominal distension with hepatomegaly and accumulation of ascitic fluid. Gross cardiomegaly is demonstrable on X-rays (Atwell and Kelly 1980; Robinson et al. 1980) and fibrosis, especially of the left ventricle, is seen at autopsy.

The primary lesion of the myocardium is caused by lytic viral replication which results in necrosis of the myofibres. This is accompanied by an inflammatory response causing infiltration of mononuclear cells into the interstitium (Robinson et al. 1980). Basophilic inclusion bodies can be demonstrated in myocardial nuclei in the acute and subacute stages. This is not possible during chronic disease when it appears the dogs have survived the phase of primary necrosis, only to develop extensive myocardial fibrosis and later die of congestive heart failure (Atwell and Kelly 1980).

Electron microscopic and immunofluorescent studies with myocardial nuclei from dogs that had died from CPV-induced myocarditis (Robinson et al. 1980; Carpenter et al. 1980) have revealed ultrastructural changes that are strikingly similar to the cytopathic effects observed in cultures of CPV-infected DKSV (dog kidney SV40 transformed) cells (Paradiso et al. 1982).

Much of the pathogenesis of CPV myocarditis is not fully understood, although the dependency of the virus on active host DNA synthesis has been well established (Kollek et al. 1982). Between 0 and 9 days of age, only 2 %–4 % of myocardial cells are dividing, and this may correspond with the most active phase of myocardial necrosis. However, in the developing foetus many other tissues are in active growth, yet a generalised infection is not seen. Experimentally induced myocarditis has been achieved in puppies that were infected *in utero* 8 days pre-partum (Lenghaus et al. 1980). If infected post-partum, then usually the enteric form of the disease develops (Robinson et al. 1980). However, Meunier et al. (1984) induced myocarditis in seronegative 5-day-old pups by oral or intraperitoneal inoculation of CPV (at 10^5 $TCID_{50}$). The resulting infection was subclinical, although histological lesions were found and virus-infected myocytes were demonstrated.

Generalised infection is unusual but has been reported in one litter of 3- to 10-day-old pups that were dying with CPV-induced pneumonitis, hepatitis, gastritis, nephritis and enteritis (Lenghaus and Studdert 1982). This has not been reported in older puppies and may suggest that congenital infection results in myocarditis because the myocardium, unlike other foetal organs, is unable to undergo substantial repair. This situation is comparable with the ataxic syndrome seen in prenatal infection of kittens with FPV (Csizar et al. 1971b) (see Fig. 1).

Lenghaus et al. (1980) have suggested that the apparent predilection for myocytes in neonatally infected pups may be explained by "permissive differentiation." In beagle pups myocytes are essentially mononuclear at birth, but from 2–8 weeks post-partum, there is a gradual transition to the binuclear state (Bishop and Hiner 1975). The pups in Lenghaus's study were experimentally infected *in utero* and, although they appeared clinically normal at birth, they had very high antibody titres on day 1, indicating that they had mounted an immune response prior to birth. This antibody level was too high to have been due solely to maternal antibodies. The pups started to develop symptoms after 2 weeks, and these could be attributed to myocarditis. This led Lenghaus to conclude that CPV or its viral DNA had remained "dormant" and protected from antibody attack in myocyte nuclei until after birth when transition to the binucleate state permitted virus multiplication.

Tattersall (1978) investigated susceptibility as a function of host cell differentiation in studies with a particular isolate of MVM. He compared infection of non-differentiating cells with cells of the same line that were induced to differentiate. Even though both types had receptors for the virus, multiplication was blocked in the non-differentiating cells. More recently, it has been confirmed with different isolates of MVM that replication requires certain tissue-specific factors expressed by cells of a particular differentiated phenotype (Guetta et al. 1986). MVM may not be unusual in this respect, for factors other than dependence on nuclear division have also been described for other parvoviruses, including H-1 and Kilham rat parvovirus (Lipton and Johnson 1983). If differentiating cells similarly provide a particularly favourable environment for CPV replication, then this could explain both the absence of active necrosis of the myocardium after 8 weeks of age, and the age limitations for susceptibility to myocardial infection. However, when seronegative pups of 3–4 weeks of age are experimentally infected, they develop the enteric form of the disease (Lenghaus et al. 1980; Robinson et al. 1980). At this age pups are still feeding from the dam, and so the mitotic index of gut epithelium is very low (Koldovsky et al. 1966). Virus should therefore preferentially localise in heart tissue at this stage.

It is generally accepted that parvoviruses only multiply in dividing cells in which replication is thought to depend upon a requirement for cellular DNA polymerases, particularly polymerases α and γ (Kollek et al. 1982). It is of interest that Lenghaus et al. (1985) have reported that FPV can replicate in cells in which cellular DNA synthesis is blocked. These observations, considered together with the close study of the pathogenesis of CPV, suggest that it is obviously an oversimplification to associate sites of virus multiplication with mitotic index alone.

2.2 Canine Parvovirus Enteric Disease

In weaned pups and non-immune older dogs, oral or nasal infection results in enteritis. The morbidity and mortality rates vary according to the age of the animal, pups of 6–20 weeks being the most susceptible. The disease is described as haemorrhagic enteritis which is accompanied by a leucopenia. Enteric lesions and changes in the number and type of circulating white blood cells resemble those symptoms produced in cats by FPV infection (Carlson and Scott 1977) and for this reason CPV has also been called canine panleucopenia by some workers (see. Fig. 1).

According to its severity, the disease is described as 'mild', 'acute' or 'peracute', when it is fatal. Hoffman and von Pock (1981) examined 111 cases, of which 63 % were between 7 and 23 weeks old. Of the total, 30 % suffered the mild form, 63 % the acute form and 7 % the peracute form. Stann et al. (1984) compared CPV enteritis in 40 'pound-source' dogs (procured from a commercial vendor) whose previous history was unknown. Clinical signs were uniformly severe, with a rapid onset of disease which resulted in death or else required euthanasia.

Kramer et al. (1980) have reported on the frequency with which different symptoms are observed, with depression and anorexia in 48 % of cases, vomiting in 85 %, diarrhoea in 100 % (but only haemorrhagic in 55 %), pyrexia in 45 %, leucopenia in 43 % and dehydration in 43 %. In addition, these authors noted encephalitis, pancreatitis and cardiopulmonary signs with a frequency of 5 %. In peracute infection, rapid dehydration and death follow within hours; in milder disease, recovery may occur within a week of onset (Kramer et al. 1980; Hoffman and von Pock 1981; McCandlish et al. 1981).

Studies such as those of Macartney et al. (1984) and Carman and Povey (1985) have characterised the pathological features of CPV enteric disease by following the course of infection in dogs orally infected with virus. These workers have shown how the sequential development of lesions in myeloid, lymphoid and intestinal tissues are related to the clinical picture. Although the disease presents clinically as an enteritis, it would appear that the lymphocyte is the primary target cell. Following oronasal infection, there is extensive replication in the lymphoid tissue of the oropharynx and mesenteric lymph nodes, from which virus is shed and carried in the blood. Virus multiplication then continues in other lymphoid tissues such as thymus and spleenic white pulp. Lytic damage is reflected in the characteristic leucopenia seen during early infection. A marked viraemic stage follows at 4–5 days post-infection, when virus can be isolated at lower titres from most other tissues and in serum, although it is not cell associated. Gross and microscopic lesions involving the upper small intestine first appear at this stage, and the degree of damage to the crypt/villus architecture subsequently dictates the clinical severity of enteric disease. In fatal cases there is complete breakdown of normal architecture and considerable necrosis of epithelium. The neutropenia seen in severe cases may be attributable to increased loss of neutrophils through damaged intestinal walls.

The role of the lymphocyte as a primary target for virus infection has been inferred from experimental studies which show that CPV inoculated by routes other than by mouth produces milder symptoms than when given orally. This presumably indicates the importance of initial virus multiplication in oropharyngeal and mesenteric lymph nodes (Pollock 1982; Potgieter et al. 1981; O'Sullivan et al. 1984).

Pathological changes are only seen in tissues which have relatively high mitotic indices, and this may be used to correlate susceptibility to host cell DNA synthesis. This is a critical factor in determining the severity of infection in gut epithelium. In cats (and rats) it is known that cellular proliferation rates are greater in the upper small intestine than in the lower regions (Carlson and Scott 1977). This is presumed to be the case for other mammals and would be in agreement with the finding that duodenal and jejunal lesions are more advanced and involve larger areas of tissue than those of the ileum. Other factors such as age, diet and intestinal flora also influence the course of enteric disease.

The mitotic index of gut epithelium is very low in suckling animals but increases sharply on weaning (Koldovsky et al. 1966). Weaning pups are therefore more susceptible to enteric infection, whereas perinatal infection results in myocarditis. As the animal ages, there is a decrease in cell turnover of gut epithelium (Thrasher and Greulich 1964a, b) and this could explain why infection in adult dogs is frequently less severe and often asymptomatic. It is of interest that difficulties encountered earlier when experimentally infecting adult dogs with CPV were overcome by oral inoculation following a period of starvation (Carman and Povey 1982). A possible explanation is that starvation causes a reduction in cell turnover, whilst re-feeding is accompanied by a surge of mitotic activity (Aldewachi et al. 1975). Inoculation of gnotobiotic and specific pathogen-free dogs rarely causes development of enteric symptoms and may only lead to seroconversion and some degree of leucopenia. Similar signs of infection are also seen in specific pathogen-free cats challenged with FPV (Carlson et al. 1977).

The presence of other gut pathogens may also influence the outcome of CPV infection. For example, a heavy round worm burden, Giardia salmonellae (Prange et al. 1982) and E. Coli infections (Isogai et al. 1989), and recent canine corona virus infection (Appel 1988) can increase the severity of CPV disease. Such pathogens seem to enhance susceptibility to CPV by stimulating cellular proliferation in gut epithelia providing sites for virus replication.

2.3 Confirmation of Canine Parvovirus Infection

The serological techniques that have been most commonly used for routine CPV diagnosis are haemagglutination (HA) activity for virus in faecal samples and haemagglutination-inhibition (HI) for sera (Walker et al. 1979; Carmichael et al. 1980; McCandlish et al. 1981; Senda et al. 1986). In the enteric form of the disease, virions can only be detected in faeces during a period of peak virus shed, and this may not coincide with maximum circulating antibody (Fig. 3).

For the sensitive identification of CPV-specific antibodies in stored sera, HI methods may not yield reproducible results, especially if comparisons are made between different laboratories. One of the

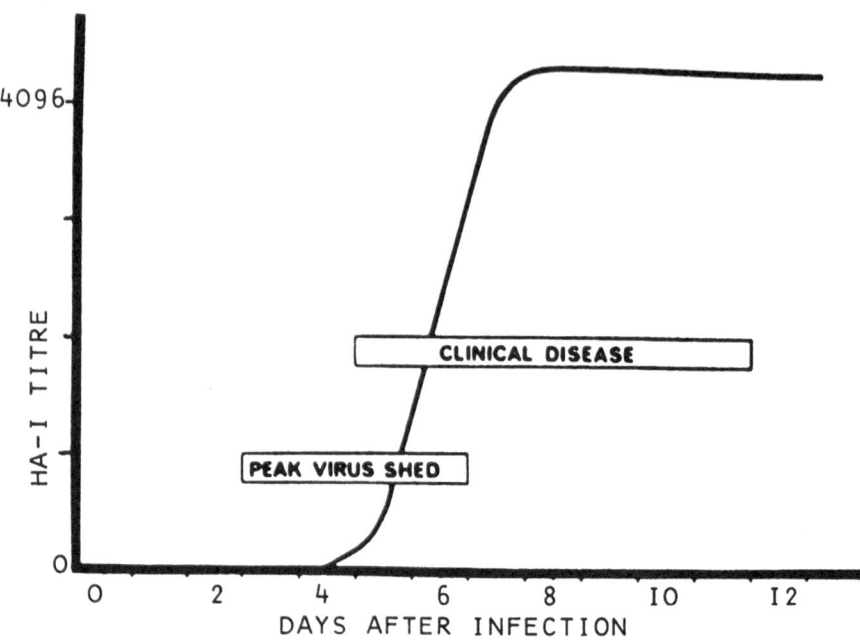

Fig. 3. Canine parvovirus enteritis – virus shedding and antibody response in relation to clinical signs (McCandlish et al. 1981)

reasons for such discrepancies seems to be that conditions for the HI test are not generally standardised, although the use of a standard CPV antiserum improves reproducibility between laboratories (Luff et al. 1987). For the sensitive detection of CPV-specific antibodies, serum-neutralisation tests are at least an order of magnitude more sensitive than HI tests, but obviously such assays require more specialised laboratory facilities. For rapid diagnosis of CPV infection, the assay of serum IgM is now probably the most useful measure. Stann et al. (1984) have made a study of 80 dogs with nonparvoviral enteritis and detected serum CPV-specific IgG antibodies (at titres of >1:25) in 85 % of them, whereas none had IgM antibodies against CPV, thus indicating the value of the IgM assay for diagnosis. Enzyme-linked immunosorbent assays (ELISA) have been developed for the class-specific measurement of IgM (Florent 1986). The presence of IgM is diagnostic of primary infection and allows differentiation between maternally derived antibody, vaccine-stimulated antibody, past infections and reinfection. Other ELISA test kits have also become available for the detection of viral antigen in faecal samples and similarly provide a rapid means of detecting CPV in practice (Herbst et al. 1986). Bartkoski et al. (1988) describe the use of a commercial ELISA kit which utilises CPV-specific monoclonal antibodies with which both pre- and post-1980 isolates can be detected and Rimmelzwann et al. (1990) and Burtonboy et al. (1991) report that ELISAs are more sensitive than either HA or HI.

3 Transmission

There are two major routes of transmission: the faecal-oral route and the transplacental route. During acute disease as many as 10^9 TCID$_{50}$ virus particles are shed per gram of faeces (Pollock 1982), and these may be passed 2–3 days post-infection, before symptoms become apparent (see Fig. 3). The minimum infectious dose for wild strains of CPV is unclear, although experience with an attenuated strain suggests that it may be a very small dose, since vaccination with this strain is accomplished by intramuscular inoculation with 16 TCID$_{50}$ (Carmichael et al. 1983). The scale of virus shed during infection and the intrinsic stability of CPV particles must have contributed to the enzootic status of the virus. Mechanical (passive) transmission of CPV is also important: infective virus may be carried out on dogs' hair and feet, on clothing and footware of dog owners, on fomites (such as feeding dishes) and by insects, notably flies. In the past this means of transmission may have been particularly significant at dog shows and in kennels where large numbers of dogs were gathered together. It must have contributed significantly to the high mortality rates recorded for young pups in the original pandemic.

In the non-immune pregnant bitch virus is transmitted vertically. Predilection for the foetus *in utero* is more widespread in parvoviruses than in any other virus group, and this is well known with MVM (Kilham and Margolis 1971), FPV (Csizar et al. 1971c), PPV (Mengeling et al. 1980) and H-1, a rodent parvovirus (Ferm and Kilham 1964). This may appear to be an efficient means of transmission for, with each infected pregnant bitch, the virus has the potential of infecting several offspring *in utero*. However, the majority of infected pups so born die of severe heart disease, and because the virus localises in myocardial nuclei, it will not be shed.

Virions may also be shed to the environment by asymptomatic carriers, and Appel et al. (1980b) have reported transmission of CPV to contact-controlled dogs by a persistent carrier.

4 Vaccination Strategies

Humoral immunity is the major component in resistance to CPV infection, and therefore prophylactic vaccination is the most effective means of control. Carmichael et al. (1981) characterised the ideal vaccine as one that should be (a) safe, (b) engender an early and vigorous immune response that endures, and (c) provide an immune barrier that interrupts spread of virulent virus. Whilst the development of homologous CPV vaccines was awaited, FPV vaccines prepared for use in cats were used with dogs. Owing to cross-antigenicity between FPV and CPV, FPV vaccines stimulated the production of antibodies in dogs which neutralised CPV particles on challenge infection. Unfortunately, such protection was short-lived and did not provide the desired immune barrier. The different types of vaccines that have been used in CPV prophylaxis are listed in Table 1 (see page 18).

It is often difficult to evaluate the results of field trials with vaccines developed for use against CPV. As has been noted by Pollock and Parrish (1985), confounding factors include the difficulty of obtaining sufficiently large numbers of participating animals, and of demonstrating the effectiveness of a vaccine against a virus which often produces mild or asymptomatic infections. Case histories for experimental animals are frequently not available, and so trials could well involve animals already naturally exposed to CPV. Comparisons of vaccines of different antigenic strengths, poor experimental designs and the use of differing criteria for evidence of "immunity" have added to the confusion.

4.1 Vaccination Failure I

A significant number of vaccination failures were reported with the early use of FPV vaccines for the protection of dogs. Gordon and Rogers (1982) evaluated the modified live FPV vaccination programme. In their study pet dogs from mixed environments were used. All were accompanied by case and vaccination (A-FPV) histories, and all were due to receive a 6-monthly booster. Faecal and serum samples were collected immediately before and 2 weeks after booster vaccination, and these were examined for HA activity (faeces) and HI titre (serum). For both pre- and post-booster faecal samples, HA activity was either negative or so low as to be insignificant, indicating that the dogs were not infected at this time, nor was sheeding of vaccine virus detectable. Pre-booster serum samples showed that for the majority of dogs HI titres were low, with 23 dogs (50 %) at <1 : 80, the minimum protective titre (Pollock and Carmichael 1982). However, for 12 dogs, titres of ≤1 : 640 were high enough to be indicative of recent asymptomatic infection. These results confirmed that A-FPV vaccine stimulated antibody production, but that protective titres were not maintained for more than 6 months. Moreover, following booster vaccination, antibody titres still remained low, and in approximately one third of cases in this study dogs did not respond to the booster.

Other studies have also revealed a wide range in vaccine-stimulated HI titres. Following a single dose of A-FPV, up to 42 % of dogs fail to develop protective titres (>1 : 80), but a second dose 2–3 weeks later may increase efficacy to 85 % (Appel et al. 1980a).

One of the reasons for early vaccination failures with all FPV vaccines was insufficient antigenic mass. FPV vaccines were prepared for use in cats and contained the optimum for them. For dogs, this had to be increased 1000 times (Appel et al. 1980a). Such novel FPV preparations were subsequently marketed as heterologous CPV vaccines for use in dogs and apparently gave protection for about 6–12 months. Nevertheless, the performance of three A-FPV vaccines licenced for use in dogs was recently investigated (Thompson et al. 1988). Vaccination was begun at 12 or 16 weeks of age, yet in both groups 67 % of dogs failed to develop protective antibody levels (≤32).

4.2 Homologous Canine Parvovirus Vaccines

CPV proved difficult to attenuate in culture and the first homologous vaccines were of the inactivated (I-CPV) type with added adjuvants.

Eugster (1980) prepared a formalin-inactivated vaccine from a virulent clinical isolate of CPV which was propagated in CRFK (Crandel feline kidney) cells. This vaccine was tested with mixed breeds of dogs between 4 and 12 weeks of age. None had a history of CPV enteritis or any other disease. These dogs received 2 ml vaccine (10^6 TCID$_{50}$/ml) intramuscularly and responded with a four-fold or greater rise in HI titre. None of the dogs challenged with virulent CPV 14 days after vaccination became clinically ill. The total white cell count changed very little in such animals, although some did shed virus with faeces.

Table 1. Summary of the different vaccines used against canine parvovirus

Heterologous feline panleucopenia (FPV)-Derived vaccines for use in dogs

Vaccine type	Efficacy and safety	Comments
Inactivated I-FPV	Safe in dogs of all ages and in the pregnant bitch; antibody levels obtained following primary vaccination (two subcutaneous injections separated by 2–3 weeks) low and short-lived	2–3-monthly boosters required to provide durable immunity, provided hypersensitivity was not evoked; did not provide an immune barrier interrupting virus spread
Modified live/ attenuated A-FPV	Safe in dogs of all ages; engendered an earlier, higher, long-lived antibody response following primary vaccination (two doses separated by 3–4 weeks)	6–12-monthly boosters required according to risk (clean or CPV-enzootic environment); not recommended in the pregnant bitch where there may be a small risk of placental passage of modified live virus; protective to individual dogs; did not provide an immune barrier
'Inactivated I-CPV' (FPV vaccine strain with antigenic mass increased by 1000)	Safe in dogs of all ages and in the pregnant bitch; initial antibody response following primary vaccination (two doses) 4–10 times greater than for I-FPV but duration of protection uncertain with considerable variation between dogs	6–12-monthly boosters required according to risk; little, if any, advantage over A-FPV but useful to immunise bitches during pregnancy so increasing maternally derived antibody titres passed to the pup *in utero* (10%) and neonatally in colostrum (90%); protective to the individual dog but unlikely to provide an immune barrier.

Homologous canine parvovirus vaccines

Vaccine type	Efficacy and safety	Recommended vaccination schedule	Comments
Inactivated I-CPV	Safe in dogs of all ages and the pregnant bitch; protective antibody response with a duration of about 12 months	Isolate pup on weaning; commence primary vaccination at 10–12 weeks with a second dose at 12–14 weeks and a booster at 5 months; annual boosters needed	Provides individual protection with some interruption of virus transmission; bitches should be given a booster before or during early pregnancy
Modified live/ attenuated A-CPV	Safe in dogs of all ages; rapid, protective antibody response is engendered and endures for up to 2 years	Isolate pup on weaning; delay vaccination until 12 weeks of age with a second dose at 15–16 weeks; booster at 5 months; then 1–2-yearly according to risk	Not recommended in pregnant bitches even though no evidence of foetal infection; bitches should be booster-injected before mating; less sensitive to interference by maternally derived antibody in pup; gives good individual protection and an adequate immune barrier

Homologous I-CPV vaccines are still readily available. However, the antigenic mass is critical with such vaccines, and in some cases the antibody response evoked may be inadequate. The use of adjuvants may improve the efficacy of I-CPV vaccines, although the choice of adjuvant may also be important. Aluminium hydroxide gel has been used as an adjuvant with some success. In a recent study, Smith and Johnson (1986) monitored the efficacy of such an I-CPV vaccine containing Alhydrogel (aluminium hydroxide) as adjuvant. This vaccine induced an antibody response within 2 days and anamnestic responses within 24 h. Except where vaccination was unsuccessful due to interference of maternally derived antibody, a single vaccination appeared to give protection for life.

Nevertheless, there is some doubt as to the value of I-CPV vaccines, for although they may afford protection for several months to individual dogs, it seems unlikely that they interrupt CPV transmission, since there is usually limited shedding of virulent virus on challenge infection.

Modified live or attenuated CPV (A-CPV) vaccines became commercially available in 1984. The difficulty with the development of A-CPV vaccines was simply that the virus resisted attenuation during serial passage in cell culture. Furthermore, CPV was difficult to handle in laboratories because it tended to escape and infect experimental dogs.

Carmichael et al. (1981) reported on the successful attenuation of a strain of CPV (C-780916) which had resisted attenuation for up to 80 serial passages in dog kidney cells. On subsequent passage at suboptimal temperatures, there was a reduction in virulence of this strain towards dogs and this was associated with a change in plaque characteristics in the cell line A-72. In this cell line the variant gave predominantly large plaques, rather than the smaller type with wild type strains in culture. This resistance to attenuation was quite unlike FPV, which can be attenuated for cats after less than 10 serial passages (Gorham et al. 1965). The modified CPV vaccine strain so developed retained both its avirulence and novel plaque characteristics on repeated passage, showing it to be genetically stable, and the novel plaque morphology proved to be a useful in vitro marker for its differentiation from wild type strains.

The safety and efficacy of this vaccine was rigorously evaluated in trials (Carmichael et al. 1983) in which more than 2000 dogs participated and 5865 doses of vaccine were used. Some of the results from these trials are presented in Table 2. Group I and II dogs neither showed symptoms of disease nor shed virus on challenge with virulent

Table 2. Duration of immunity following vaccination with attenuated canine parvovirus (Carmichael et al. 1983)

| Dog[a] | Vaccine dose (log$_{10}$ TCID$_{50}$) | Pre-vaccination | HI antibody titer Post-vaccination month | | | | | | Post-challenge (PC) results | | |
			1	4	8	12	20	24	HI titer 2 weeks	Signs[c]	Faecal virus (days PC)
Group I						Challenge[b]					
V80	6.2	<10	5120	2560	–	640			640	no	no
V81	6.2	<10	1280	1280	–	640			640	no	no
V82	6.2	<10	1280	1280	–	640			1280	no	no
V83	6.2	<10	1280	1280	–	320			640	no	no
C60	none	<10	< 10	< 10	–	< 10			5120	yes	yes (3–8)
C61	none	<10	< 10	< 10	–	< 10			5120	yes	yes (2–8)
Group II						Challenge					
V916	6.0	<10	> 5120	5120	5120	2560			2560	no	no
V917	6.0	<10	5120	2560	5120	2560			2560	no	no
V918	6.0	<10	5120	2560	5120	2560			1280	no	no
C919	none	<10	< 10	< 10	< 10	< 10			5120	yes	yes (2–7)
Group III								Challenge			
V68	5.2	<10	2560	1280	1280	640	1280	640	1280	no	no
V69	5.2	<10	10240	1280	640	1280	640	640	1280	no	no
V70	5.2	<10	2560	1280	640	1280	640	640	1280	no	no
V71	5.2	<10	1280	640	320	320	320	80	>10240	no	no
V72	5.2	<10	2560	640	640	640	640	320	640	no	no
V73	5.2	<10	640	320	320	320	320	160	320	no	yes (4–7)
C94	none	–	–	–	< 10	< 10	< 10	< 10	>10240	no	no
C95	none	–	–	–	< 10	< 10	< 10	< 10	>10240	no	yes (5–7)

[a] Vaccinated (V) or contact (C) dog introduced at time indicated. Dogs were vaccinated subcutaneously

[b] Challenge virus given by the oral/nasal route

[c] Signs of illness were anorexia, relative lymphopenia, or elevated temperatures on post-challenge days 4–6 (range). There were no enteric signs

CPV. Similar results were obtained with Group III dogs challenged 2 years after vaccination. All these dogs were maintained in isolation during the trials, and so immunity could not have been boosted by unplanned challenge exposure.

This modified live CPV vaccine was safe in dogs of all ages, and in the pregnant bitch there was no evidence of foetal infection. It gave rapid protection, and the antibody levels achieved were high and long-lived; it appeared to simulate natural immunity and interrupt virus transmission.

4.3 Vaccination Failure II

A second cause of vaccination failure was reported by Pollock and Carmichael (1982) who investigated maternally derived immunity to CPV. About 90 % of maternal antibody is passed from the dam to her pups in colostrum, and transfer is proportional to both the dam's serum antibody titre and her litter size. In general, post-suckle titres of pups reach at least 50 % of that in the dam, though levels vary between individuals in a litter according to the amount of suckling. Maternally derived antibody declines exponentially, with a half-life of 9.7 days. This is longer than with some other canine diseases, such as canine distemper virus (CDV), which has a half-life of 8.4 days (Gillespie et al. 1958).

Pollock and Carmichael (1982) established for CPV that an HI titre of $\geq 1 : 80$ was required to protect from virulent virus, whilst $> 1 : 10$ prevented successful vaccination. However, virus neutralisation and plaque reduction tests have been found to be more sensitive assays than HI for determining serum CPV-specific antibody levels. In this study, pups born of immune dams were refractory to vaccination until 14–16 weeks of age, by which time most primary vaccination schedules would normally have been completed. Consequently, maternally derived immunity was a significant cause of vaccination failure. The answer was not to be found by simply delaying vaccination because in some pups maternal immunity is lost by 6 weeks of age. There is a critical period of 2–5 weeks when maternal immunity is not high enough to protect the pup from infection but will prevent a protective vaccination response. A practical consequence of the vaccination schedules in use at this time appeared to be that approximately one third of vaccinated pups were still susceptible to enteric infection if exposed to CPV.

A similar problem with maternal immunity and vaccination against CDV has been solved by nomograph prediction. This allows the response to distemper vaccination in the pup to be predicted according to the dam's antibody titre, the age of the pup and the half-life of the antibody. Nomographs can provide the small animal clinician with a working model for vaccination. Unfortunately, two factors have hindered reliable nomograph prediction for CPV: serological assays have not been fully standardised, and there is widespread variation in maternally derived antibody levels amongst litter mates (Kukedi and Bartha 1986). Consequently, it is important to vaccinate early enough to optimise protection, but late enough to avoid interference by maternal antibodies. Vaccination success in pups is illustrated in Table 3, which compares the likely rates of immunisation achieved with CPV and CDV vaccines. However, even after vaccination, about 25 %–30 % of pups of critical ages still remain unprotected against CPV (Whur 1985; O'Brien et al. 1986; Smith and Johnson 1986).

Table 3. Likely immunisation rate in random populations of puppies vaccinated with distemper and parvovirus vaccines (Smith Kline 1984)

Age at vaccination	Percentage immunised	
(weeks)	Distemper	Parvovirus
6	50	25
9	75	35
12	95	50–60
15	99	75–90
18	99 (+)	95

4.4 Current Problems with Vaccination

Passive or maternally transferred antibody continues to be the most important factor associated with vaccination failure in the pup (Macartney et al. 1988; Burtonboy et al. 1991). Despite reports that A-CPV vaccines are protective in up to 95 % of animals whose maternal CPV-specific antibody titre is low ($\leq 1 : 10$ by HI), numerous veterinary surgeons are still finding significant failure rates.

Whur (1985) found that one third of his vaccinated pups remained unprotected and only acquired low HI titres of $1 : 32$ at best ($1 : 80$ is considered to be the minimum protective titre). In this case, as with many other small animal practices, the CPV-specific antibody titre was measured before and after primary vaccination in order to optimise schedules, yet pups still failed to mount a protective response. This may suggest that interference by maternal antibody is not the sole cause of vaccination failure. However, recently Burton-boy et al. (1991) discussed the efficacy of A-CPV vaccines in such pups, which depends on several factors including the infectious virus titre, the antigenic mass, the degree of attenuation and the vaccine strain used. In this study the serological response was shown to be proportional to the vaccine virus titre used. At a dose of $10^{7.0}$ TCID$_{50}$ an experimental vaccine induced seroconversion rates of 95, 89, 82 and 44% in pups with HI antibody titres of ≤ 8, 16, 32 and > 32 respectively. Fears that antigenic variation between pre-1980 isolates of CPV and CPV-2a could be responsible for vaccination failure have proved to be groundless since cross-protection studies have shown that vaccines developed from early isolates do protect against CPV-2a (Bartkoski et al. 1988; Appel and Carmichael 1987; Parrish et al. 1988a).

The increased use of laboratories for the evaluation of individual serum samples has also served to highlight the continued controversy over the sensitivity and value of HI and serum neutralisation techniques (McVicar 1985; Luff and Wood 1985) for the detection of CPV-specific antibodies. In the case of rubella, for example, explanations for vaccination failure have also included the limited sensitivity and/or reproducibility of the HI assay (O'Shea et al. 1981). It is reported (Anonymous 1985) that the Veterinary Products Committee (VPC) in the United Kingdom has expressed concern over the frequency of vaccination failures in pups inoculated against CPV. They have asked vaccine manufacturers to provide an explanation for this problem. The VPC does recognise that many such cases are caused by interference of passive immunity and so stress the need to rationalise vaccination schedules and develop standard sensitive serological technique for the identification of "problem" pups. Vaccine manufacturers are still confronted with these three problem areas, namely optimising vaccination schedules, the value of HI and SN test, and maternally derived immune interference.

Therefore, at present, for the home-based pet dog that can be isolated from other animals, it appears that vaccination against CPV should be delayed until 9–12 weeks of age, with a second injection 3 weeks later and a booster at 5 months. But for the high-risk pup, vaccination must begin as early as 6 weeks of age and be repeated at 3-weekly intervals until 5 months. Further booster vaccination is normally recommended for all dogs at 1- to 2-yearly intervals, according to risk (see Table 1).

5 Origins of Canine Parvovirus

The origin of CPV, more than a decade after its sudden appearance, still remains a mystery. Many hypothesis have been offered (Siegl 1984). Most are based upon CPV arising as a variant of a pre-existing parvovirus that has adapted to canines, either under field conditions or as a result of the selective pressure of attenuation, such as in the production of live vaccines. These hypotheses include:

1. Recombination of canine adenovirus with Binn's minute virus of canines
2. Variation of MEV or an MEV vaccine strain
3. Variation of a field, vaccine or laboratory strain FPV
4. Derivation from an unidentified sylvatic reservoir

Of these, the most popular is that CPV originated from FPV, and it is now widely accepted that CPV is indeed a variant of the cat virus rather than a true canine virus (Berns 1984). The autonomous parvoviruses have been characterised by their narrow host range, but FPV has proven to be an exception. The recent pandemic of CPV enteritis and myocarditis is probably the third example of serious disease in immunologically naive animal populations being caused by a virus which is closely related to FPV. RPV and MEV are both considered to have arisen as host variants of FPV before problems with CPV became apparent (see Sect 1.1).

FPV, RPV, MEV and CPV all have similar virion morphology, physico-chemical properties and pattern of enteric disease in their own hosts. All four are closely related antigenically. These features are all consistent with FPV being the common ancestor. Although there is some cross-infection and cross-protection between CPV, FPV, MEV and possibly RPV, it is important to note that these viruses normally only cause virulent disease in their natural host (Kikuth et al. 1940; Siegl 1976; Goto et al. 1984; Appel as cited by Siegl 1984; Parrish 1984). Dogs that have been experimentally infected with FPV may develop antibody against both FPV and CPV yet they show few clinical symptoms (Goto et al. 1984; Nakanishi et al. 1988).

5.1 Comparisons of Canine Parvovirus with Known Related Parvoviruses

Early evidence suggested that both RPV (Waller 1940) and MEV (Gorham et al. 1965; Johnson 1967) were serologically indistinguishable from FPV. However, using techniques currently available, existing strains of RPV, MEV and CPV can be unambiguously distinguished from FPV. Of these, the most valuable methods are serological analysis using monoclonal antibodies, viral genome mapping with restriction endonuclease enzymes and DNA sequence analysis. Nevertheless, useful information has been gained by the use of longer established techniques. CPV can be differentiated from related parvoviruses according to host cell range (Appel et al. 1980c), by virus neutralisation (VN) tests (Lenghaus and Studdert 1980), by HA activity (Carmichael et al. 1980), by HI and by precipitin reactions in gels (Flower et al. 1986).

5.1.1 Monoclonal Antibody Techniques

Monoclonal antibodies (mAb) raised against a purified pre-1980 CPV isolate and FPV-b virions were used in early studies on antigenic variation among CPV and the related parvoviruses, as shown on Fig. 4 (Parrish et al. 1982; Parrish and Carmichael 1983).

In this analysis antigenic differences were found in the virus capsids of the four related parvoviruses. Antigenic variation within a parvovirus species was also revealed. Of 13 mAbs raised against CPV, eight bound equally well to FPV, RPV and certain MEV isolates (type 1, see below), showing the presence of common antigenic determinants, whilst five were CPV-specific. CPV-e differed from the four other CPV isolates in reaction to two of the mAbs (C and D). RPV differed from FPV and MEV (type 1) at a single epitope revealed by mAb H. Antigenic variation was greatest among MEV isolates, of which three types were found. Types 1 and 2 were more closely related to FPV than type 3, and represent MEV isolates circulating in Europe and America between 1950 and 1980, whereas the two type 3 isolates were obtained in 1980 from the mid-western states of the United States of America. Antigenic differences in the MEV isolates were thought to be the consequence of very few mutations. Antigenic variation was not found between the FPV vaccine strain (a) and its natural isolates (b and c), nor among the RPV isolates, although the

Virus Isolates

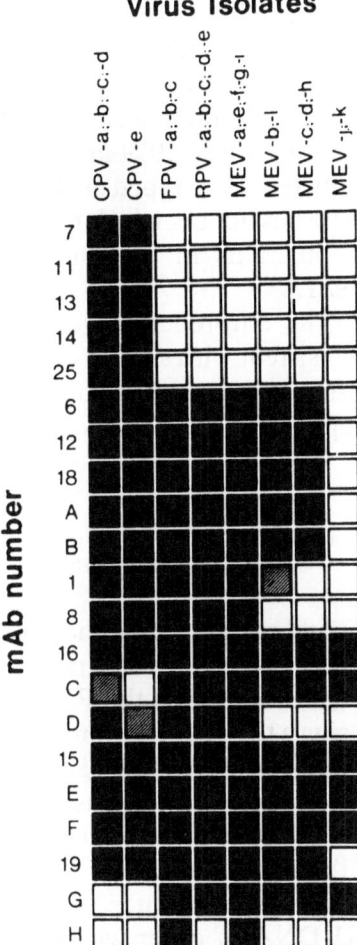

Fig. 4. Reactions with monoclonal antibodies (mAb). Reactions if isolates of CPV, FPV, RPV and MEV when titrated against all mAb using the HI test. *Shaded,* ± 1/4 the titre with the homologous virus; *crosshatched,* <1/4 but >1/20 the titre with the homologous virus; *open,* <1/20 the titre with the homologous virus, or no detectable reaction (Parrish and Carmichael 1983)

latter may have been derived from a single source (Nettles et al. 1980).

In a later study, differences were found between CPV isolates collected in the United States of America over the years 1978–1984. Two antigenic types were recognised, pre-1980 (original) CPV isolates and post-1980 (CPV-2a) isolates (Parrish et al. 1985). A selected panel of mAbs was subsequently used to further investigate this phenomenon (Parrish et al. 1988a). More than 400 CPV isolates collected from several countries across the world were typed. In addition, serum from wild coyotes was examined to determine the natural occurrence of CPV types in unvaccinated canidae (most vaccines were derived from original CPV strains).

The mAbs used were CPV/CPV-2a-specific (mAb7), CPV-specific (mAbs D and J), CPV-2a-specific (mAb IDI), CPV/FPV/MEV reacting (mAb F), and FPV-specific (mAb H). This study revealed a rapid global replacement of the original virus by CPV-2a in domestic dogs between 1979 and 1982. By this time, of 97 isolates collected, only five were typed as the original virus. Since 1983 the original CPV type has not been isolated. Similarly, CPV-2a replaced CPV in coyote populations in the United States of America between 1980 and 1981.

The differences revealed by mAbs in virion capsid proteins are consistent with studies on the structure of the genome of parvoviruses using restriction endonuclease mapping and sequence analysis of replicative form DNA.

5.1.2 Restriction Endonuclease Analysis and Sequence Analysis of Parvovirus DNA

Double-stranded replicative form (RF) parvovirus DNA can be analysed using restriction endonuclease enzymes. McMaster et al. (1981) compared the RF DNA of CPV (pre-1980 isolate) and MEV, using a total of 25 different enzymes to produce restriction enzyme site maps for the DNA of each one; of these 25, 17 nucleases cut the RF DNA of both viruses. In this way, 79 sites were mapped, 68 of which were common to both viruses (Fig. 5a). In a subsequent study (Tratschin et al. 1982), the seven enzymes which had picked out differences between CPV and MEV were used to compare the RF DNA of other wild-type and vaccine strains of MEV, FPV and CPV (Fig. 5b). The seven enzymes which performed complete restriction were *Alu*I, *Hae*III, *Hinc*II, *Hinf*I, *Hph*I, *Mbo*I and *Mbo*II. Using such selective means of analysis, the genomes of FPV and MEV differed at only a single site and so provided the FPV/MEV reference map against which other wild-type isolates and vaccine strains could be compared. Maps of the genomes of these related parvoviruses were established using the seven different enzymes.

For wild-type isolates of CPV were compared with the FPV/MEV reference. It was found that whilst CPV is closely related to the reference (with which it shares some 80 % of the sites tested), it is also clearly distinct in that if differs by eight or nine of these restriction sites. Of these differences, three *Hinf*I sites (absent in the reference) are particularly characteristic of the RF DNA of CPV. Among the four CPV isolates collected from different areas of the world, only a single difference was detected by these enzymes. The German CPV isolate had an additional MboI site at map position 76 which was not present in the Swiss, Belgian or American CPV isolates. This MboI site is present in all the FPV and MEV isolates examined in the study. The fact that the maps for three of the CPV isolates were identical when analysed using these restriction enzymes does imply that all strains of CPV stem from a common ancestor.

Wild-type MEV isolates V_{22} (which has been experimentally transmitted to dogs by Moraillon et al. 1980) and V_1 were also found to differ from the FPV/MEV reference map. These two isolates did not carry the HaeIII site at map unit 98 but did have an additional HinfI site at map unit 37, like CPV. However, it is the HaeIII site that distinguishes between MEV and FPV in the FPV/MEV reference map, and the MEV genome normally carries this site. Thus, in this respect, MEV isolates V_{22} and V_1 resemble FPV.

Six attenuated vaccine strains (five FPV derived and one MEV derived) were also compared with the FPV/MEV reference map. Of the FPV strains, A and C differed from the reference only in the length of DNA fragments at their 5' ends. This could be accounted for by the insertion, somehow, of 100 nucleotides into the viral genome. However, in FPV strain F, two extra HinfI sites were found at map units 37 and 87, in addition to an insertion of about 50 nucleotides at the 5' end. For FPV strain E, three additional HinfI sites were found at map units 37, 49 and 87. It is not only of interest that HinfI sites have been generated through the process of virus attentuation in the FPV live vaccine strains at comparable sites, but also that one of these pentanucleotide sites appears to have been generated under field conditions at map unit 37 in MEV V_{22} and V_1 (Fig. 5b). FPV vaccine strain B showed more fundamental differences from the FPV/MEV reference: it was apparently identical from map units 0–70, but the remainder of the genome showed no homology. The only MEV vaccine strain examined (MEV-D) was found to be identical to the FPV/MEV reference map as far as restriction sites were concerned.

The wild-type isolates analysed in this study had probably undergone a large number of animal to animal passages in the wild before isolation in cell culture. Tratschin and his colleagues (1982) concluded that their enzyme recognition data indicated the genomes of FPV, MEV and CPV to be stable under field conditions. Furthermore, they suggest that the multiple genetic changes like those distinguishing CPV and FPV/MEV may have developed under selective growth conditions, for example, in the course of deliberate or accidental

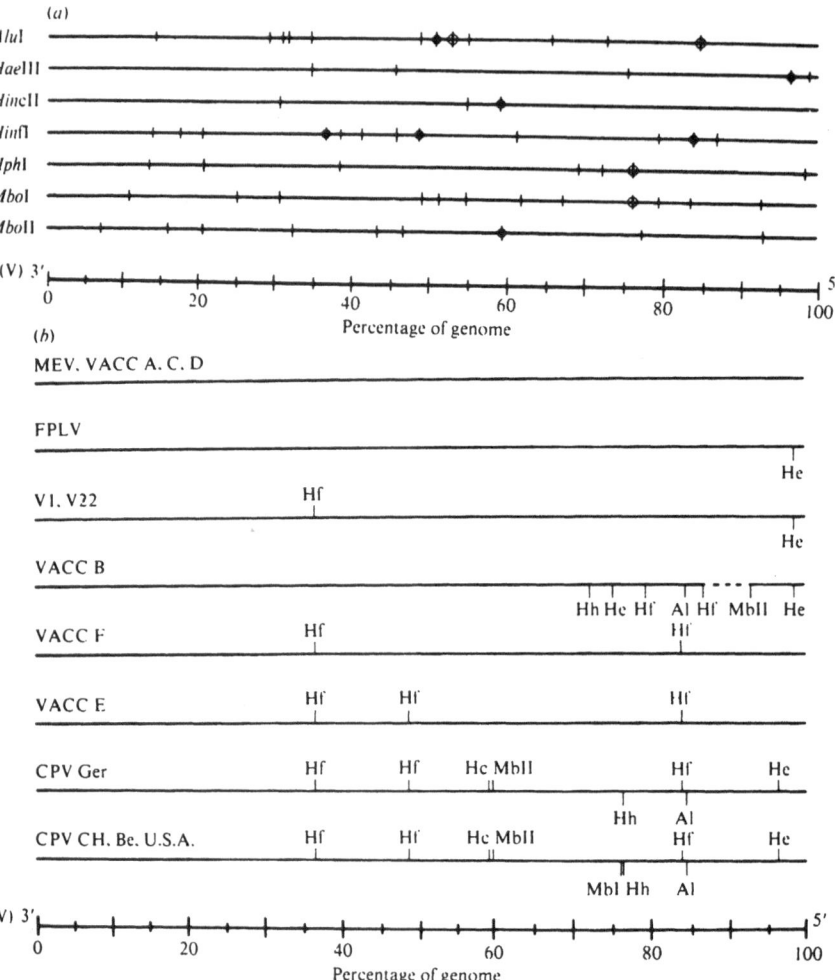

Fig. 5a, b. Restriction enzyme maps. *a* Restriction enzyme maps of mink enteritis virus (MEV) and canine parvovirus (CPV) (Swiss isolate) replicate form DNAs. The maps represent the single-stranded virus DNA. The *scale* is given in map units (percentage of the genome) starting at the virus 3'-end. *Vertical lines* indicate sites present in both MEV and CPV; ⊕, sites present in MEV; ●, sites present only in CPV (McMaster et al. 1981). *b* Comparative restriction enzyme maps of the DNAs of CPV, FPV, MEV and six vaccine strains. Each *horizontal line* represents one virus genome (single-stranded) with its 3'-end at the *left*. The *scale* is given in map units. The map of MEV is used as a basis for the comparison, and only sites which differ from those of MEV are indicated. Additional restriction enzyme sites are drawn *above* the line, whereas those missing are drawn *below* the line. Restriction sites in common for all viruses are not shown; however, they can be found in *a* as *vertical lines*. The *dotted line* at the right-hand end of the map of vaccine strain B indicates a deletion of about 0.2 kilobases. *Hf,* HinfI; *He, Hae*III; *Hh, Hph*I; *Al, Alu*I; *MbII, Mbo*II; *Hc, Hinc*II; *MbI, Mbo*I (Tratschin et al. 1982)

adaptation of FPV strains to multiply in canine cells. More recently, DNA sequence analysis has shown that specific host range, HA and neutralisation epitopes map along with antigenic variant sequences in the capsid protein gene of CPV and the related parvoviruses. Differences between CPV, CPV-2a and FPV resulted from up to three nucleotide differences (three predicted amino acid changes) between map units 59 and 64 (Vp1/Vp2) in the capsid protein gene (Parrish et al. 1988b). Thus it would appear that very few changes in the amino acid sequence of the capsid proteins would be required to alter the specific host range of these parvoviruses.

5.2 Theories and Ideas on the Origins of Canine Parvovirus

5.2.1 Recombination of Canine Adenovirus with Binn's Minute Virus of Canines

Evermann (1981) has proposed that Binn's isolate of MVC is a defective parvovirus (as adeno-associated virus) and has somehow undergone recombination with canine adenovirus type 2 (CAV-2). Such adenoviruses are known to provide 'helper' functions on co-infection, thereby permitting replication of a defective parvovirus. R. H. Johnson (as cited by Siegl 1984) considers that MVC may indeed be a defective parvovirus. This could explain the failure of MVC to propagate in any other than the Walter Reed canine cell line, especially if the cell line were contaminated with an unknown helper virus.

The fact that CAV-2 live vaccines (used against canine infectious laryngotracheitis) came into widespread use about 2 years before the appearance of CPV lends some support to Evermann's theory. However, it is difficult to test Johnson's hypothesis because it seems that even autonomous parvovirus replication can be stimulated by co-infection with an adenovirus helper (Brailovsky and Chany 1965). The replication of adeno-associated viruses is totally dependent upon function(s) provided by the adenovirus helper. However, the DNA of two such viruses do not normally hybridise and do not have any homologous regions, which one might expect to be necessary for recombination. The facts that MVC and CPV differ in their antigenic and genomic properties (Macartney et al. 1988) and that the autonomous nature of the CPV genome is well established now puts Evermann's hypothesis and others like it (e.g. Danson 1981) into disfavour (e.g. Carmichael and Binn 1981).

5.2.2 Canine Parvovirus as a Variant of Field Strains of Mink Enteritus Virus or Feline Panleucopenia Virus

Moraillon et al. (1980) proposed that CPV arose from MEV on the evidence that dogs fed on MEV-infected mink intestinal tissue developed a fatal peracute disease that was indistinguishable from CPV-induced enteritis. However, this virus, MEV-V$_{22}$, could only be subsequently propagated in a feline lung cell line, unlike CPV which has a wide host cell range in vitro. The restriction enzyme map for MEV-V$_{22}$ has been shown to be consistent with the FPV/MEV reference map (Tratschin et al. 1982) apart from two site changes: the addition of a *Hin*fI site and the loss of a *Hae*III site. MEV is normally considered to be avirulent in dogs, but in the case of strain V$_{22}$ the mapped differences in the genome could be linked to the apparent ability to infect dogs. Among the related parvoviruses virulence in dogs is otherwise confined to CPV.

The possibility that CPV arose directly from a field isolate of FPV has generally been considered unlikely. FPV is normally avirulent in dogs (Siegl 1976) and even after vaccination with an heterologous vaccine strain the multiplication of FPV is very limited and of short duration in dogs (Pollock and Carmichael 1983; Parrish 1984). Furthermore, if FPV could cause even asymptomatic infection of dogs, then it would be expected that some dogs, at least, could acquire cross-protection against wild strains of CPV, since CPV and FPV are antigenically closely related. This has not been the case, even though FPV has been enzootic in cat populations all over the world for many years.

According to the early work of Tratschin and co-workers (1982) field strains of both FPV and MEV would require at least 9–11 genetic changes to create a genome with the characteristics of CPV. These workers considered that the genomes of all the parvoviruses are genetically stable. If this is the case then it would seem unlikely that the selection pressure in the field would be sufficient to enable FPV, or MEV, to jump the species barrier into dogs.

Nevertheless, FPV has probably mutated under field conditions twice before, creating the variants RPV and MEV, both of which arose before vaccines against FPV were introduced. MEV appeared suddenly in 1947 in much the same way as CPV (Schofield 1949). The extent of alteration to the DNA of existing field strains of FPV to allow it to jump the species barrier into mink could not be determined at that time. Since then MEV could well have undergone further mutational change through selective in vivo passage. However, analysis of the viral DNA from existing field isolates with a carefully selected set of restriction enzymes reveals only a single site difference between MEV and FPV, and the two genomes are considered to be very similar (Tratschin et al. 1982). With these same nuclease digestions, CPV has been shown to have far greater differences to the DNA of FPV.

It is of interest that FPV can be transmitted to neonatal ferrets (Johnson et al. 1967) although it is avirulent in adults. This may indicate that little mutational change was necessary for FPV to adapt to the mustilidae.

Whilst there is little doubt that MEV arose as a variant from field strains of FPV, there is no real evidence for the origin of CPV from either FPV or MEV under field conditions. In the case of MEV, there is more natural variation amongst field strains, as can be demonstrated with mAb. However, restriction endonuclease analysis does not show close similarities between the viral genomes of field strains of MEV and CPV, other than a *Hin*fI site at map position 37, and this may have some evolutionary or epidemiological significance. The best available evidence to support the direct genesis of CPV from MEV is provided by the work of Moraillon and his co-workers (1980), who claim to have caused disease in dogs with MEV V_{22}. Apparently MEV isolates V_1 and V_{22} have identical restriction enzyme maps, and the pathogenicity in dogs of both of these strains would seem to warrant further investigation. However, in their recent phylogenic analysis, Parrish et al. (1988b) suggest that CPV could have arisen from an as yet unidentified FPV, RPV or MEV strain. The DNA sequences of most of the VP-1 gene and the entire VP-2 gene of 10 viruses was compared. Although FPV, MEV and RPV isolates were all closely related phylogenically, CPV isolates formed a separate cluster, which does not seem to support the notion that CPV arose as a variant of a field isolate of any of these viruses.

5.2.3 Canine Parvovirus as a Variant of a Vaccine Strain of Feline Panleucopenia Virus

It has been suggested that CPV may have arisen by mutation from a live vaccine strain of FPV (e.g. Johnson and Spradbrow 1979). Wilson (1980) reasoned that:

1. The multifocal appearance of CPV over a very short period of time could be explained if the virus were distributed in a vaccine or veterinary product.

2. The mutation of a previously existing parvovirus which enabled it to jump the pecies barrier under field conditions was unlikely.
3. Wild-type FPV cannot replicate in canine cells.
4. Modified live FPV vaccine virus may be capable of replication in dogs, and the selection pressure of virus attenuation could be sufficient to cause adaptation of FPV to canine cell lines.

Such a hypothesis is supported by the fact that FPV is easily attenuated on less than 10 serial passages (Gorham et al. 1965). If CPV arose in this way, then what is the evidence for such an adaptation?

Restriction enzyme analysis has been performed on only six vaccine strains, five of FPV and one of MEV origin. Tratschin et al. (1982) proposed that changes found in the lengths of DNA fragments from the 5' ends of the genomes (encoding capsid proteins) would affect the biological properties of the virus. Subsequent work (Parrish and Carmichael 1986) with CPV established that the ability of this virus to infect dogs can be attributed to the structure of the virion capsid proteins. In this study a mutant of CPV (CPV-102/10) was passaged in a feline cell line (NLFK) and lost its ability to multiply in canine cell line A72. It was far less infectious in dogs than the parent strain and could be distinghished by five different mAbs. However, it could not replicate in cats. Analysis of seven recombinant viruses constructed in vitro between CPV 102/10 and a wild-type strain showed that both host range and antigenic specificity mapped in the same part of the VP1/VP2 coding region. Similar studies with the four related parvoviruses and CPV-2a mapped virus neutralisation, HA activity and part of the host range epitope between mu 59 and 64 (Parrish et al. 1988b). FPV, CPV and CPV-2a differed by up to three nucleotides only. Thus small genetic changes involving as few as three amino acids may be all that is required for FPV to adapt to canine cells.

It is also significant that two out of five FPV vaccine strains examined carried additional HinfI sites at map positions 37 and 87. Restriction enzymes cutting at the HinfI site recognise a pentanucleotide sequence (G $\downarrow$ ANTC) and three such sites are characteristic of the CPV genome. Considering the very small size of the parvovirus genome (approximately 5000 nucleotides), it would seem unlikely that three HinfI sites could evolve under field conditions, yet with the FPV vaccine strains two or three such sites have been generated by the selective pressure of virus attenuation. However, it should be noted that the two vaccine strains (E and F, Fig. 5b) that were most similar to CPV (pre-1980) have now been sequenced (Parrish et al. 1988b) and it is surprising that these do not appear to be any more closely related to CPV than wild-type FPV strains.

Adaption of FPV to a canine cell line could possibly have been a consequence of a deliberate attempt to speed up attenuation of a wild-type virus or a laboratory strain. Other parvoviruses have been adapted to novel cell lines which do not normally support their replication. For example, Cartwright et al. (1969) successfully adapted porcine parvovirus to growth in a variety of human cell lines, and these may be considered now as biological variants (Hallauer et al. 1971). High passage of parvoviruses does induce changes in virus host range specificity and antigenic characteristics as has been shown with MVM (Astell et al. 1986), H-1 (Rhodes 1978a, b) and CPV (Parrish and Carmichael 1986). Reed et al. (1988) sequenced the RF DNA of a high passage isolate of CPV (CPV-N) and compared it with standard genomes of CPV and FPV. It may be relevant that homology between FPV and CPV-N was greater than that between CPV-N and CPV isolate b. The nucleotide homologies of the capsid genes between CPV-N and FPV were 98% and 99%, respectively. Alternatively, adaptation to canine cells could have been accidental in a laboratory that handled several veterinary products. Latency and persistent infection are common aspects of host/parvovirus associations and cell lines are known to have been developed from latently infected tissues (e.g. Siegl 1976; O'Reilly and Whitaker 1969). Recently, Metcalf et al. (1989) have discussed such problems with tissues derived from neonatal rabbits infected with lapine parvovirus (LPV). Parvoviruses have also apparently been introduced into cell lines by chance and remained undetected through several passages (e.g. Hallauer et al. 1971; Nettleton and Rweyemamu 1980).

5.2.4 Derivation from a Sylvatic Reservoir

Testing of stored canine sera in the United States of America, Canada, Australia, New Zeland, South Africa and Europe showed that CPV-specific antibodies (by HI test) were not present for about 10 years before the 1980 CPV pandemic (Carmichael et al. 1980; Walker et al. 1980; Helfer-Baker et al. 1980). In Europe retesting of stored sera did reveal such CPV antibodies in Belgium as early as 1976 (Schwers et al. 1979) and in France in 1977 (Petermann and Chapuis 1981). Of the Belgium samples, 3 of 56 gave very high HI titres ($\leq 1:1280$). However, the three positive samples were all derived from apparently healthy adult dogs from a single source. The French samples, like the Belgian ones, had been collected randomly from seemingly healthy dogs and revealed a distinct pattern of seroconversion. Samples collected in the years 1975–1976 were all clearly negative, but in 1977 20 % of samples were seropositive, and this proportion rose to 36 % in 1978. Such evidence seems to suggest that asymptomatic CPV infection was common in France by 1977 and yet overt disease was not recorded until October 1979, some 2 years later. Interestingly, the first documented case of CPV disease in France was reported in Paris in an adult bitch that had just returned from a dog show in Belgium (Petermann and Chappuis 1981). Further conflicting evidence comes from Greece (Koptopoulos et al. 1986) with a report that three dog sera from a group of 28 collected in 1974 around the Thessaloniki area were apparently seropositive for CPV (by HI test), yet there were no signs of any disease being caused by this virus at the time.

The panzootic of CPV disease around 1980 was characterised by its highly virulent nature, and the veterinary profession is of the general opinion that overt disease could not have passed unnoticed in France during 1977, and so the idea of an unidentified sylvatic reservoir of CPV arose. However, no such reservoir has yet been identified. Schwers et al. (1983) looked at the fox (another member of the canidae) as a possible sylvatic reservoir in France. They found that only 2.8 % of wild foxes tested had serum antibody against CPV. Clearly the fox acted neither as a reservoir nor as a vector of CPV infection in Europe. In the United States of America, Thomas et al. (1984) followed the course of a natural asymptomatic CPV infection in three separate wild coyote populations. Seroprevalence in these coyotes was comparable with that seen in domestic dog populations. In these animals CPV-specific antibodies were not found before 1979 and no evidence was found that coyotes could constitute a natural source of CPV infection. However, it may be significant that in a more recent study of sera of wolves from Minnesota, CPV-specific antibodies were first detected by HI in samples taken in 1976, before or at about the same time that this virus first began infecting dog populations in the United States of America (Goyal et al. 1986).

5.2.5 An Attempt to Correlate the Evidence

It is difficult to explain the almost simultaneous appearance of CPV in the United States of America and Australia, other than by distribution in a veterinary product. The virus could have been carried around the world by infected dogs and on dog owner's clothing. However, many countries require a health certificate from a local veterinary inspector in the country of origin and compulsory quarantine measures may be imposed on imported animals, as in the United Kingdom. If CPV had been spread from country to country solely via infected dogs, then reports of CPV disease in quarantine kennels would have been expected. Likewise, if carriage was via contaminated dog owners' clothing, then isolated CPV infection among pet dogs would have been noted well before the disease became enzootic. At present the most satisfactory explanation for the spread of the original isolate across international borders seems to be through a contaminated veterinary product.

The appearance of antibodies against CPV in stored canine sera in France might be used to support the notion that CPV or its precursor arose from a veterinary product, in particular a live vaccine strain of FPV. It is postulated that such a strain could have carried the three additional *Hin*fI sites, characteristic of the CPV genome, that have also been found in some of the vaccine strains of FPV. If some of this modified live virus escaped complete attenuation or reacquired partial virulence in some way, and was used in FPV cat vaccination programmes in France during the period up to 1977, then such a virus could be shed with cat faeces and picked up by dogs, accounting for the antibodies first seen in canine sera in 1977. In these dogs the virus might only be capable of limited replication at this stage, causing an asymptomatic infection. However, before termination by the immune response mounted in such dogs, there could be some faecal shedding of the virus. Repeated passage of the virus through dogs, by faecal-oral transmission, could then follow and favour further adaptation to canine cells without manifestation of any serious symptoms. If dog tissues with such silent infection were now used for the preparation of laboratory cell lines, then the CPV precursor could contaminate any veterinary product derived from such cells or cross-contaminate other clean cell lines. By such means the virus could be transmitted over a wide geographical area and find its way into immunologically naive dogs in other countries in which it could replicate and further adapt in virulence sufficiently to cause overt disease. The adapted virus could now represent CPV, and on reintroduction into France give the outbreaks first recorded in 1979.

This interpretation is entirely speculative, but could account for the early seroconversion of asymptomatically infected dogs in France, whilst being in agreement with other ideas on the sudden appearance and rapid pandemic spread of CPV. However, it is not apparently consistent with the Belgian evidence. It should be noted that only a very low frequency (3 of 56) of positive sera was detected in the early Belgian samples. The analysis of Greek canine sera from 1974 (Koptopoulos et al. 1986) similarly revealed seroconversion at a low rate (3 of 28 samples). However, in the Greek study, titres by serum neutralisation (SN) tended to be lower than HI titres, and this is a reversal of the usual pattern for CPV. Assuming that the SN test is more specific, then this could point to some antigenic diversity in these cases. These authors therefore argue that the three dogs in question were not infected with CPV, but some other related parvovirus that was not transmissible from dog to dog. This raises the possibility that the Belgian workers may similarly have been measuring cross-reactivity of a related parvovirus other than CPV.

The only novel point in this interpretation is the proposal that CPV has evolved from an FPV vaccine strain by undergoing reversion to virulence on passage through cats and subsequently dogs. It is well established that vaccine strains of viruses can, and do, revert to virulence both in vivo and in vitro (Clark 1978). Repeated in vivo passage can lead to increased virulence with parvovirus (Moore and Nicastri 1965). This phenomenon is also illustrated by live CDV (Rockborn strain), which can show increased virulence after as few as 6 serial passages in dogs or 10 passages in vitro (Appel 1978). Furthermore, the shedding of attenuated poliovirus with faeces does provide a means for the infection of susceptible contacts in which varying degrees of reversion to virulence may result in manifestation of type 2 or type 3 poliomyelitis (Kew et al. 1981). In this example, the number of individuals so infected by polio is very small because the majority of the population commonly have the benefit of herd immunity. However, with foot and mouth disease, the introduction of residual live virus into non-immune cattle through vaccines has probably led to serious outbreaks of the disease crossing national boundaries (Brooksby as quoted by Cherfas 1982).

The interpretation presented relies upon the evolution of the CPV precursor, in France, in dogs that developed a protective immune response, so preventing fulminant disease symptoms and maintaining the asymptomatic carrier state. It is suggested that it was the export of this partially adapted virus to areas in which dogs had not been exposed that facilitated the final transition to the canidae during the critical period 1977–1978.

The concept of CPV arising as a variant of FPV replicating in and adapting to dogs has recently been considered by Reed et al. (1980).

5.2.6 Divergence of Two Canine Parvovirus Strains

The global replacement of the initially successful virus by CPV-2a over the years 1979–1982 poses yet another puzzling problem. In the United States of America the new variant appeared in 1979, when one of six isolates obtained from domestic dogs was typed as CPV-2a. Within a year, an approximately equal proportion of old and new viruses were collected, and by 1981 the new virus predominated. CPV-2a also replaced the old isolate in three wild (unvaccinated) coyote populations in the United States of America between 1980 and 1981. In Denmark the new variant predominated by 1980, when 121 of the 125 isolates examined typed as CPV-2a. Very few French, Belgian, Japanese and Australian isolates were examined but, from the available data, the new isolate appeared in Japan (1 of 4) in 1979 and in France (1 of 1) in 1980. Australian isolates collected in 1978/1979 were of the old type and, although 1980 isolates were not examined, in 1981 the single isolate collected was of the new type (Parrish et al. 1988a). These workers pointed out that dogs that had recovered from CPV infection with pre-1980 isolate were immune to infection by CPV-2a so that the new virus spread in populations that should have been immune. In addition, CPV-2a could not have spread to Europe, Japan or Australia in live CPV/FPV vaccines since these were not used in these countries during the early years.

Based on studies with mAbs and phylogenic analysis, CPV and CPV-2a diverged prior to 1978 according to Parrish et al. (1988a). This would imply that CPV-2a spread through the world in much the same way as the original virus. However, it seems highly unlikely that the two viruses diverged *prior to* the original pandemic because CPV and CPV-2a do not appear to differ in their transmissibily and pathogenicity in dogs. Moreover, infection with either one of the isolates stimulates protective immunity against the other and vaccines based on the original isolate protect from infection by CPV-2a. It is difficult to explain the selective advantage for either the initial success of the original virus or its replacement by CPV-2a. It would seem far more likely that CPV-2a evolved from CPV during natural passage through canines and evidence in support of this idea has been provided by Senda et al. (1988). These workers examined HA activity and antigenicity, using mAb, in 30 Japanese virus isolates collected from dogs between 1979 and 1985, and compared them with two American reference strains. According to these criteria the viruses could be divided into two groups. Group 1 comprised the early isolates which resembled FPV/MEV antigenically and in their temperature-sensitive HA activity. In group 2, comprising isolates collected in the later years of the study including current CPV isolates, HA activity was temperature-independent, and this group differed antigenically. Further evidence to suggest that these changes occurred on natural passage through dogs was inferred from isolates collected during the transitional period. These resembled FPV/MEV antigenically whilst showing different HA activity. Parrish et al. (1988c), describe a non-haemagglutinating mutant of pre-1980 CPV that arose naturally after high passage through feline kidney cell line (NLFK). Although antigenically indistinguishable from wild-type virus, this mutant replicated well in both canine and feline cell lines with no reduction in virus titre.

6 Parvoviruses and Human Disease

In recent years autonomous parvoviruses have been implicated in an expanding spectrum of human disease. However, there is no evidence, that any of the parvoviruses infecting humans is antigenically related in any way to the CPV, FPV, MEV group (Siegl 1976).

The best characterised human parvoviruses, B19, previously known as serum parvovirus-like agent, was first isolated from serum obtained from healthy human donors (Cossart et al. 1975) and subsequently from the sera of sickle cell anaemia patients during aplastic crises (Pattison et al. 1981; Serjeant et al. 1981). Characterisation of the linear single-stranded DNA and the genome organisation of this virus permits its classification as a member of the Parvoviridae (Summers et al. 1983; Cotmore and Tattersall 1984; Ozawa et al. 1988). DNA hybridisation, restriction enzyme and sequence analysis have shown that B19 is, in fact, distinct from both the autonomous parvovirus and the Dependovirus groups. B19 replication is autonomous, unlike the dependoviruses (e.g. adeno-associated virus 2) which require co-infection with a helper virus for replication. However, B19 resembles the dependoviruses in that both ends of the genome carry long inverted terminal repeat sequences (Shade et al. 1986) and it packages DNA strands of both positive and negative polarity into virions in equal numbers. One other autonomous parvovirus, LU III, may package DNA in this way under certain conditions (Muller and Siegl 1983; Berns et al. 1985).

B19 appears to be very similar to the human parvovirus (HPV) first isolated from human faeces (Paver et al. 1973; Turton et al. 1990). B19 is associated with febrile illness (Schneerson et al. 1980), fifth disease or erythema infetosium (Anderson et al. 1984) and aplastic crisis in patients with some form of haemolytic or sickle cell anaemia (Serjeant et al. 1981; Anderson et al. 1985). Fifth disease has been recognised in children for many years across the world and the characteristic sign is an erythema of the cheeks which has led to the common description "slapped-cheek syndrome". However, infection of adults is commonly complicated by arthropathies and adenopathies which may persist for some time. Antibodies against fifth disease are usually acquired during childhood and approximately 65 % of young adults are seropositive for B19 (Cohen et al. 1983; Cohen and Buckley 1988). The virus is transmissible in human clotting factor and 89 to 97 % of haemophiliac children acquire antibody to B19 (Anderson and Pattison 1984; Williams et al. 1990). However its usual route of transmission is respiratory (Anderson et al. 1985).

The predilection of B19 for actively dividing cells such as erythroid progenitor cells is an important factor in the pathogenesis of this virus. However at least two other factors are important, target cell differentiation and the immune status of the host (Pattison 1990). Thus B19 may infect different foetal tissues and infection may be persistent in the immunocompromised host, whilst the arthopathies as seen in adults may be associated with a transient autoimmune state.

The virus can cross the human placenta and infect foetal tissues causing severe anaemia leading to hydrops foetalis, intrauterine death or absorption (Brown 1984; Carrington et al. 1987; Enders & Biber 1990). B19 DNA, antigen and virions have been detected in foetal tissues (Clewley et al. 1987) and recent reports implicate the virus as a cause of congenital malformations (Van Elsacker-Niel et al. 1989; Hartwig et al. 1989).

In addition to its role in aplastic crisis in patients with some form of chronic haemolytic anaemia, B19 is now clearly associated with persistent infection in immunocompromised patients such as those undergoing chemotherapeutic or immunosuppressive therapy (Kurtzman et al. 1988; Graeve et al. 1989; Takahashi et al. 1991). Such infection may lead to chronic aplaisia of all marrow elements (Rao et al. 1990; Morinet and Perol 1990). In patients infected with the human immunodeficiency virus (HIV) B19 is now a recognised opportunistic pathogen. In addition to complicating treatment of HIV disease, patients may develop severe, chronic anaemia. Fortunately most patients respond well to intravenous immunoglobulin therapy (Chrystie et al. 1990; Mitchell et al. 1990; Bowman et al. 1990; Astro et al. 1990).

B19 is clearly a cause of persistent arthropathies and arthritis in susceptible adults, and much interest has been shown in its possible involvement in RA (rheumatoid arthritis) (Reid et al. 1985). B19 DNA has been detected in synovial fluid (Dijkmans et al. 1988). Some of these cases do fulfil the criteria for RA (Naides et al. 1990) but, on follow up, patients do not develop the joint pathology of the classical disease (see Woolf 1990). Whilst the pathogenesis of B19 associated arthropathy remains unclear, there is recent evidence to suggest that infection may be accompanied by a transient, subclinical autoimmune state (Knight and Isenberg 1990; Saski et al. 1990; Soloninka et al. 1989).

RA-1 is another candidate HPV that has been associated with arthritis. It has been possible to detect both RA-1 related antigens and viral nucleic acid sequences in clinical specimens from different patients (Simpson et al. 1984, 1985).

Our understanding of the role of parvoviruses in human disease is still incomplete. The most important factor limiting work on human parvoviruses has been the successful development of cell culture systems for the *in vitro* propagation of virus. As yet, the only such system for B19 seems to be cells in erythropoietin-stimulated human bone marrow cultures. It is likely that the human parvoviruses will present some new and challenging problems. For example, in the case of B19, the terminal inverted repeat sequences in viral DNA may well be consistent with the ability of this viral genome to integrate into the host cell chromosome under conditions in which lytic viral replication is blocked. Five groups of B19 variants have so far been described (Mori et al. 1986). Although antigenic variation has not been found, these variants differ at one or more restriction enzyme sites. Individual genomes do not appear to correlate with the different disease presentations of the virus (Mori et al. 1987) but a correlation between genome types and prevalence has been observed in Japan (Umene & Nunoue 1990).

7 Prospects and Problems

CPV appears to be the third variant of FPV to have arisen, following MEV and quite probably RPV. Although there is little doubt that

FPV is able to cross the species barrier, available evidence seems to suggest that CPV is only virulent in canines. However, in contrast to FPV, MEV and RPV, CPV does have the ability to replicate in the laboratory in a range of cultured animal cells derived from several species, including human and bovine cell lines. Not only does it replicate, but high passage of CPV in feline and canine cells can induce antigenic and genetic variants. This raises the possibility that there may be some potential for CPV to spread into further species unrelated to the dog. At present, though, there is no irrefutable evidence of CPV causing disease in any other species of animal.

Nevertheless, new infections of animals caused by parvoviruses like the FPV, MEV, CPV group are appearing from time to time. Veijalainen and co-workers (1984) isolated a parvovirus from blue foxes with a mild transient diarrhoea in Finland. Serological analysis in 1980 showed the presence of CPV-specific antibodies (HI titre 1 : 640–1 : 2560) in foxes on 13 farms, but when the virus was isolated in 1983, it resembled FPV and MEV rather than CPV in its host cell range, HA activity, and antigenic structure. More recently this isolate has been placed in the type 2 MEV group (Veijalainen 1988). In another suspected parvovirus infection of porcupines (Frelier et al. 1984), symptoms resembled those of CPV. Four out of six animals were suffering, or died, from enteritis and another had myocarditis. Histological examination showed features consistent with CPV infection, although EM, serological and virological investigations failed to detect a parvovirus or any other causative agent. In this case, the evidence linking porcupine infection to a CPV-like virus is entirely circumstantial. Another new parvovirus has been found to infect horses (Wong et al. 1985). This is the first report of an equine parvovirus and it was responsible for abortions in 16 of 70 mares on a farm in Manitoba. On isolation from two of the aborted foetuses, the virions were found to have the properties of an autonomous parvovirus, although they were not related to either CPV or FPV. Yet further reports link parvoviruses or parvovirus-like agents to psittacine beak and feather disease (Gough et al. 1989), necrotic hepatitis of rabbits (Gregg & House 1989), pneumonia in the corn snake, *Elapha guttata* (Ahne and Scheinert 1989) and infectious anaemia in chickens (Goodwin et al. 1989, 1990). Two new parvoviruses of pigs have been described (Foni et al. 1989; Yasuhara et al. 1989). The latter isolate is distinct from porcine parvovirus (PPV) and infection is now widespread in Japan. The rabbit parvovirus described by Gregg and House (1989) is thought to be responsible for a virulent disease syndrome with high (80–90%) morbidity, reported from China, Korea, Italy,

Switzerland and other European countries. In 1988 a similar disease was reported from Mexico City that rapidly spread through 13 states within a few months.

The sudden appearance of new viruses such as these may justify concern over the use of live vaccine strains for the control of animal parvovirus infections. Even though it has been shown that inactivated parvovirus vaccines are inferior, the continual use of live vaccines must increase the opportunity for selection of variants. CPV could well have arisen this way and other new parvoviruses could evolve by this means in future. Nevertheless, in the United Kingdom vaccines for veterinary use do have to comply with strict regulations (1968 Medicines Act [revised 1971], Therapeutic Order Act 1952) on safety, efficacy and quality.

The obvious hazard associated with the use of live vaccines is reversion to increased virulence. With animal parvoviruses this problem can be considered at two levels – in manufacture and in use.

Many more doses of animal vaccines are used each year than human vaccines since livestock is continually being reared for slaughter. The majority of animal vaccines are of the modified live (attenuated) type. Such vaccine strains should be genetically stable, but inevitably their development and large-scale production must carry some risk. In the case of animal parvovirus vaccine strains, manufacture provides the opportunity for reversion and, coupled with the ability to adapt to new cells lines, this could have serious consequences in unvaccinated animal populations.

In general, in the construction of live animal vaccines a compromise has to be reached such that attenuation is not so extreme as to prevent multiplication in the natural host, and so it may be necessary to tolerate mild symptoms of disease. With the animal parvoviruses this may include faecal shedding of virus, and it is known that this process can provide further opportunity for reversion.

There are a number of ways in which live vaccine strains can be monitored to guard against the possibility of reversion, for example, the selection of vaccine strains with novel plaque characteristics that are associated with a reduction in virulence for the natural host. The strain of CPV (C-780916) used in the production of Carmichael's live CPV vaccine (Carmichael et al. 1981, 1983) carries such a marker, and reduced virulence in dogs is accompanied by a change from small to large plaque morphology. Likewise, a mutant of CDV (derived from the Onderstepoort strain) is known (Cosby et al. 1983) that has novel plaque characteristics associated with reduced virulence for weaning hamsters (in this case a change from large to small plaques). However, as with human vaccine strains of viruses, the best safeguard for vaccine strains destined for use in animals is likely to be by the definition of nucleotide sequences in the viral genome so as to specify both a satisfactory level of attenuation and immunogenicity.

With the aim of improving safety, efficacy and quality, a number of new approaches to vaccine construction have been proposed. These were the subject of a Cold Spring Harbor Symposium (Chanock and Lerner 1983) and are as follows:

1. Expression of protective antigen in a prokaryotic host
2. Expression of protective antigen in a eukaryotic host
3. Synthetic peptides as immunogens
4. Antigenic mimicry with anti-idiotype antibodies
5. Attenuation by deletion mutation
6. Attenuation by gene reassortment
7. Attenuated vaccine virus as a vector for heterologous viral protective antigen
8. Selection of epitope mutants using neutralising monoclonal antibody

Strategies 1–4 describe the construction of non-replicating antigens, and the remainder involve attenuated mutants for use as live vaccines.

With the development of recombinant DNA technology, it has become possible to produce immunogenic viral antigen through the expression of the genes coding for capsid proteins in cloned host cells. The small size of the DNA genome of parvoviruses, and the simplicity of capsid construction with a maximum of just three proteins (VP1, VP2', VP2) makes viruses like CPV ideal for such genetic manipulation. Some of these novel approaches have been considered for the construction of safer CPV vaccines.

The introduction of fragments of CPV-DNA into *E. coli* using plasmid vectors has resulted in the successful production of a β-galactosidase fusion protein (Smith and Halling 1984). Although this antigen stimulated the formation of antibodies in rabbits which precipitated authentic CPV structural proteins in vitro, these antibodies did not neutralise whole CPV particles. It is possible that by varying the method of antigen preparation, the immune response can be improved to provide antibodies capable of neutralising virus infectivity. Certainly the identification of the sequences coding for coat proteins in the genomes of CPV (Rhode 1985) and FPV (Carlson et al. 1985) must facilitate the specific cloning of the coat protein genes for these viruses in either a prokaryotic or eukaryotic host. It is probable that better immunogenicity will be obtained from such proteins by the expression of genes in eukaryotic cells. However, cloned viral antigens are likely to be most useful in the control of systemic virus infections in which neutralising antibody rather than mucosal IgA antibody or cytotoxic T cells are critical. Available evidence for CPV seems to suggest that live vaccines will be superior because they evoke both tissue and humoral responses (Rice et al. 1982).

The stabilisation of live CPV vaccine strains by deletion mutation may hold some promise. Such mutants characteristically never undergo reversion, but one difficulty with such an approach with CPV vaccine strains may be that, as yet, the genes concerned with virulence have not been identified, although generally with small viruses available information indicates that every gene product must function to give virulence.

Synthetic peptide immunogens have been used with mixed success with other viruses, but normally antigenicity is poor and must be enhanced by the use of an adjuvant. However, animals inoculated with such peptides may be primed so that they then respond to subsequent inoculation with subimmunising doses of whole virus by developing protective levels of neutralising antibody. Therefore, synthetic immunogens may have some use in priming the moderate response seen with inactivated CPV vaccines.

Even now, a number of years after its sudden appearance, CPV, and subsequently CPV-2a, still remains an important pathogen of the domestic dog, although the myocardial form of the disease is now rarely seen in the United Kingdom and mortality rates for the enteric form can be reduced by rapid medication. The virus is maintained in domestic dog populations by the infection of young pups as they lose their passive maternal protection. The only real means of controlling CPV disease is by vaccination, ideally with modified live homologous vaccines which should simulate natural disease better and help to increase herd immunity. Nevertheless, there are still important advances to be made in the development of safe and effective prophylactic CPV vaccines and in the design of vaccination programmes, for at present, despite vaccination, up to one third of fully vaccinated pups remain at risk. Realistically, it seems that at best CPV, like FPV in cats, will only ever be reduced to acceptable enzootic levels.

Acknowledgements. The authors are indebted to Dr. Desmond McCarthy (Queen Mary and Westfield College) for his helpful discussions and critical reading of the manuscript. We would also like to thank Dr. Barry Bush (Royal College of Veterinary Surgeons) for his valuable help. The encouragement and guidance given by Michael Gordon, John Madley (Glaxovet Limited) and Dr. B. J. Cohen (Central Public Health Laboratory, Colindale) are also gratefully acknowledged.

References

Abinati FR, Warfield MS (1961) Recovery of a haemadsorbing virus (HADEN) from the gastrointestinal tract of calves. Virology 62:113–125

Afshar A (1981) Canine parvovirus infection – a review. Vet Bull 51 (8):605–612

Ahne W, Scheinert P (1989) Reptilian viruses: isolation of parvovirus-like particles from corn snake Elapha guttata (Colubridae). Zentralbl Veterinarmed B 36:409–412

Aldewachi HS, Wright NA, Appleton DR, Watson AJ (1975) The effect of starvation and refeeding on cell population kinetics in the rat small bowel mucosa. J Anat 119:105–121

Anderson MJ (1987) Human Parvovirus infections. J Virol Meth 17:175–181

Anderson MJ, Pattison JR (1984) The human parvovirus: brief review. Arch Virol 82:137–148

Anderson MJ, Lewis E, Kidd IM, Hall SM, Cohen BJ (1984) An outbreak of erythema infectiosum associated with human parvovirus infection. J Hyg 93:85–93

Anderson MJ, Higgins PG, Davis LR, Williams JS, Jones SE, Kidd IM, Pattison JR, Tyrrell DAJ (1985) Experimental Parvoviral infection in humans. J Inf Dis 152 (2):257–265

Anonymous (1985) "CPV vaccine failures interest the CPV." Comment. Vet Rec 177 (9):193

Appel MJG (1978) Reversion to virulence of attenuated canine distemper virus in vivo and in vitro. J Gen Virol 41:385–393

Appel MJG (1988) Does canine caronovirus augment the effect of subsequent parvovirus infection? Vet Med 83:360–366

Appel MJG, Carmichael LE (1987) Can a commercial vaccine protect pups against a recent field isolate of parvovirus? Vet Med 82 (10):1091–1093

Appel MJG, Scott FW, Carmichael LE (1979) Isolation and immunisation studies of a canine parvo-like virus from dogs with haemorrhagic enteritis. Vet Rec 105:156–159

Appel MJG, Carmichael LE, McGregor DD, Pollock RVH (1980a) Canine Parvovirus vaccination. Mod Vet Prac 61:983–985

Appel M, Meunier P, Pollock R, Glickman L, Greisen H, Carmichael LE (1980b) Canine viral enteritis. In: Mayer E (ed) Satelite symposium on diseases of small animals, Tel-Aviv, October 1980. Israel Association for Buiatrics, Haifa, pp 171–179

Appel M, Meunier P, Pollock R, Greisen H, Carmichael L, Glickman L (1980c) Canine viral enteritis, a report to practitioners. Canine Prac 7:22–36

Astell CR, Gardiner EM, Tattersall P (1986) DNA sequence of the lymphotropic variant of minute virus of mice, MVM(i), and comparison with the DNA sequence of the fibrotropic prototype strain. J Virol 57, 656–669

Atwell RB, Kelly WR (1980) Canine parvovirus: a cause of chronic myocardial fibrosis and adolescent congestive heart failure. J Small Anim Prac 21:609–620

Bachmann PA, Hoggan MD, Kurstak E, Melnick JL, Pereira HG, Tattersall P, Vago C (1979) Parvoviridae: Second Report Intervirology 11:248–254

Bartkoski MJ Jr, Curren M, Dees C, Stroh S (1988) Canine parvovirus immunodiagnosis and vaccination procedures. Comp Anim Pract 2:30–33

Berns KI (ed) (1984) The Parvoviruses. Plenum, New York

Berns KI, Adler S (1972) Separation of two types of adeno-associated virus particles containing complementary single strands. J Virol 9:394–396

Berns KI, Muzyczka N, Hauswirth WW (1985) Replication of Parvoviruses. In: Fields BN, Knipe DM, Chanock RM, Melnick JL, Roizman B, Shope RE (eds) Virology. Raven, New York, pp 415–432

Binn LN, Lazar E, Eddy GA, Kajima M (1970) Recovery and characterization of a minute virus of canines. Infect Immun 1:503–508

Bishop SP, Hine P (1975) Cardiac muscle cytoplasmic and nuclear development during canine neonatal growth. Rec Adv Stud Cardiac Struc Metab 3:637–656

Bouillant A, Hanson RP (1965) Epizootiology of mink enteritis. III. Carrier state in mink. Can J Comp Med 29:183–189

Brailovsky C, Chany C (1965) Un facteur produit pa l'adeneno-virus 12 en culture cellulaire, stimmulant la multiplication du virus k du Rat. CR Acad Sci (III) (Paris) 260:2634–2637

Brown T, Anand A, Richie LD, Clewley JP, Reid TMS (1984) Intrauterine parvovirus infection associated with hydrops fetails (letter). Lancet 2:1033–1034

Burtonboy S, Charlier P, Hertoghs J, Lobmann M, Wiseman A, Woods S (1991) Performance of high titre attenuated canine parvovirus vaccine in pups with maternally derived antibody. Vet Rec 128: Vol. 377–381

Carlson J, Rushlow K, Maxwell I, Maxwell F, Winston S, Hahn W (1985) Cloning and sequence of DNA encoding structural proteins of the autonomous parvovirus feline panleukopenia virus. J Virol 55 (3):574–582

Carlson JH, Scott FW (1977) Feline Panleukopenia (II). The relationship of intestinal mucosal cell proliferation rates to viral infection and development of lesions. Vet Path 14:173–181

Carlson JH, Scott FW, Duncan JR (1977) Feline Panleukopenia (I). Pathogenesis in germfree and specific pathogen-free cats. Vet Path 14:79–88

Carman PS, Povey RC (1985) Pathogenesis of canine parvovirus-2 in dogs. (I) Haematology serology and virus recovery. (II) Histopathology and antigen identification in tissues. Res Vet Sci 38 (2):134–150

Carman S, Povey C (1982) Successful experimental challenge of dogs with canine parvovirus-2. Can J Comp Med 46:33–38

Carmichael LE, Binn L (1981) New enteric viruses in the dog. In: Cornelius CE, Simpson CF (eds) Advances in veterinary science and comparative medicine, vol 25. Academic, New York, pp 1–37

Carmichael LE, Jaubert JC, Pollock RVH (1980) Haemagglutination by canine parvovirus serologic studies and diagnostic applications. Am J Vet Res 41:784–791

Carmichael LE, Jaubert JC, Pollock RVH (1981) A modified live canine parvovirus strain with novel plaque characteristics. Cornell Vet 71:408–427

Carmichael LE, Jaubert JC, Pollock RVH (1983) A modified-live canine parvovirus vaccine. II. Immune response. Cornell Vet 73:13–29

Carpenter JL, Roberts RM, Harpster NK, King NW Jr (1980) Intestinal and cardiopulmonary forms of parvovirus infection in a litter of pups. J Am Vet Med Ass 176:1269–1273

Carrington D, Gilmore DH, Whittle MJ, Aitken D, Gibson AAM, Patrick WJA, Brown T, Caul EO, Field AM, Clewley JP, Cohen BJ (1987) Material serum alpha-fetoprotein – a marker of fetal aplastic crisis during intrauterine human parvovirus infection. Lancet 1:433–435

Cartwright SF, Lucas M, Huck RA (1969) A small haemagglutinating porcine, DNA virus I. Isolation and properties. J Comp Pathol 79:371

Chanock RM, Lerner RA (eds) (1983) Modern approaches to vaccines: molecular and chemical basis of virulence and immunogenicity. Cold Spring Harbor, New York

Cherfas J (1982) Man Made Life. Blackwell, Oxford

Clark HF (1978) Rabies viruses increase in virulence when propagated in neuroblastomer cell cultures. Science 199:1072–1075

Clewley JP, Cohen BJ, Field AM (1987) Detection of parvovirus B19 DNA, antigen and particles in the human foetus. J Med Virol 23:367–376

Cockburn A (1947) Infectious enteritis in the Zoological Gardens, Regents Park. Br Vet J 103:261–262

Cohen BJ, Buckley MM (1988) The prevalence of antibody to human parvovirus B19 in England and Wales. J Med Microbiol 25:151–153

Cohen BJ, Mortimer PP, Pereira MS (1983) Diagnostic assays with monoclonal antibodies for the human serum parvo-like virus (SPLV). J Hyg (Camb) 91:113–130

Cosby SL, Morison J, Rima BK, Martin SJ (1983) An immunological study of infections of hamsters with large and small plaque canine distemper viruses. Arch Virol 76 (3):201–210

Cossart YE, Field AM, Cant B, Widdows D (1975) Parvovirus like particles in human sera. Lancet 1:72–73

Cotmore SF, Tattersall P (1984) Characterization and molecular cloning of a human parvovirus genome. Science 226:1161–1165

Crawford LV (1966) A minute virus of mice. Virology 29:605–612

Csiza CK, Scott FW, De Lahunta A, Gillespie JH (1971a) Immune carrier state of feline panleukopenia virus-infected cats. Am J Vet Res 32:419–426

Csiza CK, Scott FW, De Lahunta A, Gillespie JH (1971b) Pathogenesis of Feline panleukopenia virus in susceptible newborn kittens (1) Clinical signs, hematology, serology and virology. Inf Immun 3:833–837

Csiza CK, Scott FW, De Lahunta A, Gillespie JH (1971c) Feline viruses XIV: Transplacental infections in spontaneous panleukopenia of cats. Cornell Vet 61:423–439

Danson DLG (1981) Origins of parvoviral infections. J Am Vet Med Ass 178:373

Dijkmans BA, Van Elsacker-Niele AN, Salimans MN, Van Albada Kuipers GA, deVries E, Weiland HT (1988) Human Parvovirus-B19 DNA in synovial fluid. Arthritis Rheum 31:279–281

Else RW (1980) Fatal haemorrhagic enteritis in a puppy associated with a parvovirus infection. Vet Rec 106:14–15

Enders G, Biber M (1990) Parvovirus B19 infections in pregnancy. Behring Inst Mitt 85:74–78

Ernst S, Montes S, Martin R (1988) A retrospective epidemiological study of the risk factors associated with the occurrence of parvovirus infection in a canine hospital population. Arch Med Vet (Chile) 20 (1):38–43

Eugster AK (1980) Studies on canine parvovirus infections: development of an inactivated vaccine. Am J Vet Res 41:2020–2024

Eugster AK, Nairn C (1977) Diarrhoea in puppies: parvoviruslike particles demonstrated in their faeces. Southwest Vet 30:50–60

Evermann JF (1981) Origin of canine parvovirus. J Am Vet Med Ass 178:340

Evermann JF, Foreyt W, Maag-Miller L, et al. (1980) Acute haemorrhagic enteritis associated with canine coronavirus and parvovirus infections in a captive coyote population. J Am Vet Med Ass 177:784–786

Fastier LB (1968) Feline panleucopenia – a serological study. Vet Rec 83:653–655

Ferm VH, Kilham L (1964) Congenital anomalies induced in hamster embryos with H-1. Science 145:510–511

Fletcher KC, Eugster AK, Schmidt RE, Hubbard GS (1979) Parvovirus infection in maned wolves. J Am Vet Med Ass 175:897–900

Florent G (1986) Enzyme-linked immunosorbent assay for single serum diagnosis of canine parvovirus disease. Vet Rec 119:479–480

Flower RLP, Wilcox GE, Robinson WF (1980) Antigenic differences between canine parvovirus and feline panleucopenia virus. Vet Rec 107:254–256

Foni E, Gualandi GL, Capucci L (1989) Characterization of a parvovirus isolated from a pig foetus. Microbiologica 12:227–280

Frelier PF, Leininger RW, Armstrong LD, Nation PN, Povey RC (1984) Suspected parvovirus infection in porcupine. J Am Vet Med Ass 185 (11):1291–1294

Gillespie JH, Baker JA, Burgher J, Robson D, Gilman B (1958) The immune response of dogs to distemper virus. Cornell Vet 48:103–126

Glickman LT, Domanski LM, Patronek GJ, Visintainer F (1985) Breed-related risk factors for canine parvovirus enteritis. J Am Vet Med Ass 187 (6):589–594

Goodwin MA, Brown J, Miller SL, Smeltzer MA, Steffens WL, Waltman WD (1989) Infectious anemia caused by a parvovirus-like virus in Georgia broilers. Avian Dis 33:438–445

Goodwin MA, Brown J, Smeltzer MA, Crary CK, Girchik T, Miller SL, Dickson TG (1990) A survey for parvovirus-like virus (so called chick anemia agent) antibodies in broiler breeders. Avian Dis 34:704–708

Gordon JC, Angrick EJ (1986) Canine parvovirus: environmental effects on infectivity. Am J Vet Res 47 (7):1464–1467

Gordon JC, Rogers WA (1982) Field evaluation of a canine parvovirus vaccination program, using feline origin modified live virus vaccine. J Am Vet Med Ass 180 (12):1429–1431

Gorham JR, Hartsough GR, Burger D, Lust S, Sato N (1965) The preliminary use of attenuated feline panleukopaenia virus to protect cats against panleukopaenia and mink against virus enteritis. Cornell Vet 55:559–566

Goto H, Hirano T, Uchida E, Wantabe K, Shinagawa M, Ichijo S, Shimizuk K (1984) Comparative studies of physiochemical and biological properties between canine parvovirus and feline panleukopenia virus. Jap J Vet Sci 6 (4):519–526

Gough RE, Collins MS, Gresham AC (1989) A parvovirus-like agent associated with psittacine beak and feather disease. Vet Rec 125:41

Goyal SM, Mech LD, Rademacher RA, Khan MA, Seal US (1986) Antibodies against canine parvovirus in wolves of Minnesota: a serologic study from 1975 through 1985. J Am Vet Med Ass 189 (9):1092–1094

Graeve, JL, de-Alarcon PA, Naides SJ (1989) Parvovirus B19 infection in patients receiving cancer therapy: the expanding spectrum of disease. Am J Pediatr Hematol Oncol 11:441–444

Gregg DA, House C (1989) Necrotic hepatitis of rabbits in Mexico: a parvovirus. Vet Rec 125:603–604

Guetta E, Rom D, Tal J (1986) Developmental-dependent replication of minute virus of mice in differentiated mouse testicular tissue. J Gen Virol 67:2549–2554

Hallauer C, Kronauer G, Siegl G (1971) Parvoviruses as contaminants of permanent human cell lines. 1. Virus isolation from 1960–1970. Arch Ges Virusforsch 35:80–90

Hartwig NG, Vermeij-Keers C, Van Elsacker-Niele AM, Fleuren GJ (1989) Embryonic malformations in a case of intrauterine parvovirus B19 infection. Teratology 39:295–302

Hauswirth WW (1984) Autonomous parvovirus DNA structure and replication. In: Berns KI (ed) The Parvoviruses. Plenum, New York, pp 129–153

Helfer-Baker C, Evermann JF, Mckeirnan AJ, Morrison WB, Slack RL, Miller CW (1980) Serological studies on the incidence of canine enteritis viruses. Canine Pract 7:37–42

Herbst W, Danner K, Krauss H, Behrens F (1986) Detection of canine parvovirus in practice using a parvovirus ELISA test kit. Prakt Tierarzt 67(6):480–482

Herringham WP, Andrews FW (1888). Two cases of cerebellar disease in cats with staggering. St Bath's Hosp 24:112

Hitchcock LM, Scarnell J (1979) Canine parvovirus isolated in UK. Correspondence. Vet Rec 105:172

Hoffman R, Pock U von (1981) Epidemiology and symptoms of parvovirus infection in the dog. Prakt Tierarzt 62:16–23

Isogai E, Isogai H, Onuma M, Mizukoshi N, Hayashi M, Namioka S (1989) Esherichia E coli associated endotoxemia in dogs with parvovirus infection. Nippon Juigaku Zasshi 51:597–606

Johnson BJ, Castro AE (1984) Isolation of canine parvovirus from a dog brain with severe necrotising vasculitis and encephalomalacia. J Am Vet Med Ass 184:1398

Johnson RH (1967) Feline panleukopenia virus – in vitro comparison of strains with a mink enteritis virus. J Small Anim Pract 8:319–324

Johnson RH, Margolis G, Kilham L (1967) Identity of feline ataxia virus with feline panleukopenia virus. Nature 214:175–177

Johnson RH, Collings DF (1969) Experimental infection of piglets and pregnant gilts with a parvovirus. Vet Rec 85:446–447

Johnson RH, Spradbrow PB (1979) Isolation from dogs with severe enteritis of a parvovirus related to feline panleukopenia virus. Aust Vet J 55:151

Kelly WR, Atwell RB (1979) Diffuse subacute myocarditis of possible viral aetiology – a cause of sudden death in pups. Aust Vet J 55:36–37

Kew OM, Nottay BK, Hatch MH, et al. (1981) Multiple genetic changes in the oral poliovaccines upon replication in humans. J Gen Virol 56:337–347

Kikuth W, Gönnert R, Schweickert M (1940) Infektiose Aleukozytose der Katzen. Zentralbl Bakteriol 146:1–17

Kilham L, Margolis G (1971) Fetal infection of hamsters, rats and mice induced with minute virus of mice (MVM). Teratology 4:43–61

Knight B, Isenberg DA (1990) Autoantibodies in sera from patients with parvovirus B19 infection (letter) J Rheumatol 17:416–417

Koldovsky O, Sunshine P, Kretchmer N (1966) Cellular migration of intestinal epithelia in suckling and weaned rats. Nature 212:1389–1390

Kolleck R, Tseng BT, Goulain M (1982) DNA polymerase requirements for parovirus H-1 DNA replication in vitro. J Virol 41:982–989

Koptopoulos G, Papdoulos O, Papanastasopoulo M, Cornwell HJC (1986) Presence of antibody cross-reacting with canine parvovirus in the sera of dogs from Greece. Vet Rec 118:332–333

Kramer JM, Meunier PC, Pollock RVH (1980) Canine Parvovirus: update. Vet Med Small Anim Clin 75 (10):1541–1555

Kukedi A, Bartha (1986) Antibody response after vaccination of dogs possessing maternal and acitve immunity against parvovirus. Magyar Allatorvosok Lapja 41:477–480

Kurtzman GJ, Meyers P, Cohen BJ, Ammunullah A, Young NS (1988) Persistant B19 parvovirus infection as a cause of severe chronic anaemia in children with acute lymphocytic leukaemia. Lancet 2:1159–1162

Kurtzman GJ, Cohen BJ, Field AM, Oseas R, Blaese RM, Young NS (1989) Immune response to B19 parvovirus and an antibody defect in persistent viral infection. J Clin Invest 84:1114–1123

Lenghaus C, Studdert MJ (1980) Relationship of canine panleucopaenia (enteritis) and myocarditis parvoviruses to feline panleucopaenia virus. Aust Vet J 56:152–153

Lenghaus C, Studdert MJ, Finnie JN (1980) Acute and chronic canine parvovirus myocarditis following intrauterine inoculation. Aust Vet J 56:465–468

Lenghaus C, Studdert MJ (1982) Generalised parvovirus infection in neonatal pups. J Am Vet Med Assoc 181:41–45

Lenghaus C, Mun TK, Studdert MJ (1985) Feline panleukopenia virus replicates in cells in which cellular DNA synthesis is blocked. J Virol 53 (2):345–349

Lipton HL, Johnson RT (1972) The pathogenesis of rat virus infections in the new born hamster. Lab Invest 27:508–513

Luff PR, Wood GW (1985) CPV serology. Correspondence. Vet Rec 116 (21):575

Luff PR, Wood GW, Herbert CN, Thornton DH (1987) Canine parvovirus serology: a collaborative assay. Vet Rec 120 (12):270–273

Macartney L, McCandlish IAP, Thompson H, Cornwell HJC (1984) Canine parvovirus enteritis 1: Clinical, haematological and pathological features of experimental infection. Vet Rec 115:201–210

Macartney L, Parrish CR, Binn LN, Carmichael LE (1988a) Characterization of minute virus of canines (MCV) and its pathogenicity for pups. Cornell Vet 78 (2):131–145

Macartney L, Thompson H, McCandlish IAP, Cornell JHC (1988b) Canine parvovirus: interaction between passive immunity and virulent challenge. Vet Rec 122:573–576

Mann PC, Bush M, Appel MJG, Bechler BA, Montali RJ (1980) Canine parvovirus infection in South American canids. J Am Vet Med Ass 177:779–783

Mayor HD, Torikai K, Melnick J, Mandel M (1979) Plus and minus single-stranded DNA separately encapsidated in adeno-associated satellite virions. Science 166:1280–1282

McCandlish IAP, Thompson H, Cornwell HJ, Macartney L (1980) Canine parvovirus infection. Vet Rec 107:204–205

McCandlish IAP, Thompson H, Fisher EW, Cornwell HJC, Macartney L, Walton IA (1981) Canine parvovirus infection. In Pract 3 (3):5–14

McMaster GK, Tratschin JD, Siegl G (1981) Comparison of canine parvovirus with mink enteritis virus by restriction site mapping. J Virol 38:368–371

McVicar ID (1985) CPV vaccination. Correspondence. Vet Rec 117 (3):71

Mengeling WL, Paul PS, Brown TT (1980) Transplacental infection and embryonic death following maternal exposure to porcine parvovirus near the time of conception. Arch Virol 65:55–62

Metcalf JB, Lederman M, Stout ER, Bates RC (1989) Natural parvovirus infection in laboratory rabbits. Am J Vet Res 50:1048–1051

Meunier PC, Cooper BJ, Appel MJG, Slauson DO (1984) Experimental viral myocarditis: parvoviral infection of neonatal pups. Vet Path 21 (5):509–515

Mitra A, Snyder CE, Bates RC, Banerjee PT (1983) Comparative physicochemical and biological properties of two strains of kilham rat virus, a non-defective parvovirus. J Gen Virol 61:43–54

Moore AE, Nicastri AD (1965) Lethal infection and pathological findings in A × C rats inoculated with H virus and RV. J Natl Cancer Inst 35:937–944

Moraillon A, Moraillon R, Person JM, Paroda AL (1980) Parvovirose canine: L'ingestion d'organes de vision atteint d'enterite a virus decleuche chez le chien une maladie identique a la maladie spontanee. Rec Med Vet 156:539–548

Mori J, Beattie P, Melton DW, Cohen BJ, Clewley JP (1987) Structure and mapping of the DNA of human parvovirus B19. J Gen Virol 68:2797–2806

Morinet F, Perol Y (1990) B19 chronic bone marrow failure: a persistent parvovirus infection of humans. Nouv Rev Fr Hematol 32:91–94

Morinet F, Tratschin JD, Perol Y, Siegl G (1986) Comparison of 17 isolates of the human parvovirus B19 by restriction enzyme analysis. Arch Virol 90:165–172

Mortimer PP, Cohen BJ, Buckley MM, Cradock-Watson JE, Ridehalg MKS, Burkhardt F, Schilt U (1985) Human parvovirus and the foetus (letter). Lancet 2:1012

Muller DE, Siegl G (1983) Maturation of parvovirus LuIII in a subcellular system 1. Optimal conditions for in vitro synthesis and encapsidation of viral DNA. J Gen Virol 64:1043–1054

Nakanishi S, Ichijos S, Osame S, Goto H (1988) Experimental infection of conventional dogs with feline panleukopenia virus. J Jap Vet Med Ass 41:100–103

Nettles VF, Pearson JE, Gustafson GA, Blue JL (1980) Parvovirus infection in translocated raccoons. J Am Vet Med Ass 177:787–789

Nettleton PF, Rweyemamu MM (1980) The association of calf serum with the contamination of BHK 21 clone 13 suspension cells by a parvovirus serologically related to the minute virus of mice. Arch Virol 64:359–374

O'Brien SE, Roth JA, Hill BL (1986) Response of pups to modified live canine parvovirus component in a combination vaccine. J Am Vet Med Ass 188 (7):699–701

O'Reilly KJ, Whitaker AM (1969) The development of feline cell lines for the growth of feline infectious enteritis (panleukopenia) virus. J Hyg 67:115–124

O'Shea S, Parsons G, Best JM, Banatvala JE, Balfour H H Jr (1981) How well do low levels of rubella antibody protect? Correspondence. Lancet 2:1284

O'Sullivan G, Durham PJK, Smith JR, Campbell RSF (1984) Experimentally induced severe parvoviral enteritis. Aust Vet J 61:1

Ozawa K, Ayub J, Young N (1988) Translational regulation of B19 parvovirus capsid protein production by multiple upstream AUG triplets. J Biol Chem 263 (22):10922–10926

Paradiso PR, Rhodes SL, Singer II (1982) Canine parvovirus: a biochemical and ultrastructural characterization. J Gen Virol 62:113–125

Parker GA, Steadham MA, Dellen A van (1977) Myocarditis of probable viral origin in chickens. Avian Dis 21:123–132

Parrish CR (1984) Canine parvovirus and feline parvovirus: structure and function. Diss Abstr Int B 45 (5):1397

Parrish CR (1990) Emergence, natural history and variation of canine, mink and feline parvoviruses. Ad Virus Res 38:403–450

Parrish CR, Carmichael LE (1983) Antigenic structure and variation of canine parvovirus type-2, feline panleukopenia virus and mink enteritis virus. Virology 129:401–414

Parrish CR, Carmichael LE (1985) Molecular analysis of the host range of canine parvovirus. EMBO workshop on parvoviruses, Grangeneuve/Posieux, Switzerland, September 16–19, 1985

Parrish CR, Carmichael LE (1986) Characterization and recombination mapping of an antigenic and host range mutation of canine parvovirus. Virology 148:121–132

Parrish CR, Carmichael LE, Antczak DF (1982) Antigenic relationships between canine parvovirus type-2, feline panleukopenia virus and mink enteritis virus using conventional anti-sera and monoclonal antibodies. Arch Virol 72:267–278

Parrish CR, O'Connell PH, Evewrmann JF, Carmichael LE (1985) Natural variation of canine parvovirus. Science 230:1046–1048

Parrish CR, Have P, Foreyt WJ, Everymann JF, Senda M, Carmichael LE (1988a) The global spread and replacement of canine parvovirus strains. J Gen Virol 69 (5):1111–1116

Parrish CR, Aquadro CF, Carmichael LE (1988b) Canine host range and a specific Epitope map along with variant sequences in the capsid protein gene of canine parvovirus and related feline, mink and raccoon parvoviruses. Virology 166:293–307

Parrish CR, Burtonboy G, Carmichael LE (1988c) Characterization of a nonhemagglutinating mutant of canine parvovirus. Virology 163 (1):230–232

Pattison JR (1990) The pathogenesis of diseases associated with B19 virus. Behring Inst Mitt 85:55–59

Pattison JR, Jones SE, Hodgson J, Davis LR, White JM (1981) Parvovirus infections and hypoplastic crisis in sickle cell anaemia. Lancet 1:664–665

Paver WK, Caul EO, Ashley CR, Clarke SKR (1973) A small virus in human faeces. Lancet 1:237–239

Petermann HG, Chappuis G (1981) Immunoprophylaxe der Parvovirus-Infektion beim Hund. Prakt Tierarzt 62:52–58

Pintel D, Padachanji D, Astell CR, Ward DC (1983) The genome of minute virus of mice, an autonomous parvovirus, encodes two overlapping transcription units. Nucl Acids Res 11:1019–1038

Pollock RVH (1982) Experimental canine parvovirus infection in dogs. Cornell Vet 72:103

Pollock RVH, Carmichael LE (1981) Newer knowledge about canine parvovirus. In: Proceedings, 30th Gaines vet symp, pp 36–40

Pollock RVH, Carmichael LE (1982) Maternally derived immunity to canine parvovirus infection: transfer, decline and interference with vaccination. J Am Vet Med Ass 180:37–42

Pollock RVH, Carmichael LE (1983) Use of modified feline panleukopenia virus vaccine to immunize dogs against canine parvovirus. Am J Vet Res 44:169

Pollock RVH, Parrish CR (1985) Canine parvovirus. In: Olsen RG et al. (eds) Comparative pathobiology of viral diseases, vol 1. CRC Press, Boca Raton, Florida, pp 145–177

Potgieter LDN, Jones JB, Patton CS, Webb-Martin JA (1981) Experimental parvovirus infection in dogs. Can J Comp Med 45:212

Prange H, Schneider E, Schimke E, Zieger M, Grass M (1982) Clinical aspects of parvovirus enteritis in dogs. Monatshefte Veterinärmed 37:453–459

Reed AP, Jones EV, Miller TJ (1988) Nucleotide sequence and genome organisation of canine parvovirus. J Virol 62:226–276

Reid DM, Brown J, Reid TMS, Renie JAN (1985) Human parvovirus associated arthritis: a clinical and laboratory description. Lancet 1:422–425

Rhode SL III (1985) Nucleotide sequence of the coat protein gene of canine parvovirus. J Virol 54 (2):630–633

Rhode SL III (1978a) Replication processes of the parvovirus H-1. X. Isolation of a mutant defective in replicative-form DNA replication. J Virol 25:215–223

Rhode SL III (1978b) Defective interfering particles of parvovirus H-1. J Virol 27:347–356

Rice JB, Winters KA, Krakowka S, Olsen RG (1982) Comparison of systemic and local immunity in dogs with canine parvovirus gastroenteritis. Infect Immun 38 (3):1003–1009

Rimmelzwaan GF, Juntti M, Klingeborn B, Groen J, Uytdehaag FG, Osterhaus AD (1990) Evaluation of enzyme-linked immunosorbent assays based on monoclonal antibodies for the serology and antigen detection in canine parvovirus infections. Vet-Q 12:14–20

Robinson WF, Huxtable CR, Pass DA (1980) Canine parvovirus myocarditis, a morphological description of the natural disease. Vet Pathol 17:282–293

56   Vella, Ketteridge

Rogers S (1987) Canine parvovirus in Zimbabwe: incidence and control. Zimbabwe Vet J 18:34–41

Saski T, Takahashi Y, Yoshinaga K, Sugamura K, Shiraishi H (1989) An association between human parvovirus B19 infection and autoantibody production (letter) J Rheumatol 16:708–709

Schneerson JM, Mortimer PP, Vandervelde EM (1980) Febrile illness due to a parvovirus. Br Med J 280B:1580

Schofield FW (1949) Virus enteritis in mink. North Am Vet 30:651–654

Schwers A, Pastoret PP, Burtonboy G, Thiry E (1979) Frequence en Belgique de l'infection a parvovirus chez le chien, avant et apres l'observation des premiers cas cliniques. Ann Med Vet 123:561–566

Schwers A, Barrat J, Blancou J, Naenhoudt M, Pastoret PP (1983) Recherche d'anticorps envers le parvovirus canin dans les serums de renards en France. Ann Med Vet 127:544–546

Scott FW, Csiza CK, Gillespie JH (1970) Feline viruses. V. serum-neutralization test for feline panleukopenia. Cornell Vet 60:183–191

Senda M, Hirayman N, Yamamoto H, Kurata K (1986) An improved haemagglutination test for the study of canine parvovirus. Vet Microbiol 12 (1):1–6

Serjeant GR, Toplay JM, Manson K, Serjeant BE, Pattison JR, Jones SE, Mohamed R (1981) Outbreaks of aplastic crisis in sickle cell anaemia associated with a parvovirus-like agent. Lancet 2:595

Shade RO, Blundell MC, Cotmore SF, Tattersall P, Astell CR (1986) Nucleotide sequence and genome organisation of human parvovirus B19 isolated from the serum of a child during aplastic crisis. J Virol 58 (3):921–936

Siegl G (1976) The Parvoviruses. In: Garcl S, Hallauer C (eds) Virology monographs, vol 15. Springer, Berlin Heidelberg New York

Siegl G (1984) Canine parvovirus. Origin and significance of a "new" pathogen. In: Burns KI (ed) The parvoviruses. Plenum, New York, pp 363–387

Simpson RW, McGinty L, Simon L, Smith CA, Godzeski CW, Boyd RJ (1984) Association of parvoviruses with rheumatoid arthritis in humans. Science 223:1425–1428

Simpson RW, Van Leeuwen D, Zazra JJ, Smith CA, Watson KF (1985) EMBO workshop on parvoviruses, Grangeneuve/Posieux, Switzerland, September 16–19, 1985

Smith JR, Johnson RH (1986) Observations on the use of an inactivated canine parvovirus vaccine. Vet Rec 118:385–387

Smith S, Halling SM (1984) Expression of canine parvovirus-β-galactosidase fusion proteins in Escherichia coli. Gene 29:263–269

Smith Kline (1984) Canine parvovirus technical information manual. Animal Health Limited, Cavendish Road, Stevenage, Herts, UK

Soloninka CA, Anderson MJ, Laskin CA (1989) Anti-DNA and antilymphocyte antibodes during acute infection with human parvovirus B19. (letter). J Rheumatolo 16:777–781

Spalholz BA, Tattersall P (1983) Interaction of minute virus of mice with differentiated cells: strain-dependent target cell specificity is mediated by intracellular factors. J Virol 46:937–943

Stann SE, DiGiacomo RF, Giddens WE Jr, Evermann JF (1984) Clinical and pathological features of parvoviral diarrhoea in pound-source dogs. J Am Vet Med Ass 185 (6):651–655

Summers J, Jones SE, Anderson MJ (1983) Characterization of the genome of the agent of erythrocyte aplasia, permits its classification as a parvovirus. J Gen Virol 64:2527–2532

Surleraux M, Bodeus M, Delferriere N, Burtonboy G (1986) Canine parvovirus; comparative study of wild and vaccinal strains at the structural polypeptide level. Ann Med Vet 130 (2):119–123

Takahashi M, Moriyama Y, Shibata A, Takai K, Sanda M (1991) Anemia caused by parvovirus in an adult patient with acute lymphoblastic leukemia in complete remission. Eur J Haematol 46:47

Tattersall P (1978) Susceptibility to minute virus of mice as a function of hostcell differentiation. In: Ward DC, Tattersall P (eds) Replication of mammalian parvoviruses. Cold Spring Harbor, New York, pp 131–149

Tattersall P, Bratton J (1983) Reciprocal productive and restrictive virus-cell interactions of immunosuppressive and prototype strains of minute virus of mice. J Virol 46:944–955

Tattersall P, Shatkin AJ, Ward DC (1977) Sequence homology between the structural polypeptides of minute virus of mice. J Mol Biol 111:375–394

Thomas NJ, Foreyt WJ, Evermann JF, Windberg LA, Knowlton FF (1984) Seroprevalence of canine parvovirus in wild coyotes from Texas, Utah and Idaho (1972 to 1983). J Am Vet Med Ass 185 (11):1283–1287

Thompson H, McCandlish IAP, Cornell HJC, Macartney L (1985) Parvovirus and reduced fertility: no Link. Correspondence. Vet Rec 116:378

Thompson H, McCandlish IAP, Cornwell HJC, Macartney L, Maxwell NS, Weipers AF, Willis IRW, Black JAC, Mackenzie AC (1988) Studies of parvovirus vaccination in the dog: the performance of live attenuated feline parvovirus vaccines. Vet Rec 122 (16):378–385

Thomson GW, Gagnon AN (1978) Canine gastroenteritis associated with a parvovirus like agent. Can Vet J 19:346

Thrasher JD, Greulich RC (1964a) The duodenal progenitor population. I. Age related increase in the duration of the cryptal progenitor cycle. J Exp Zoo 159:39–46

Trasher JD, Greulich RC (1964b) The duodenal progenitor population II. Age related changes in size and distribution. J Exp Zoo 159:385–396

Tratschin JD, McMaster G, Kronauer G, Siegl G (1982) Canine parvovirus: Relationship to wild-type and vaccine strains of feline panleukopenia virus and mink enteritis. J Gen Virol 61:33–41

Thurn J (1988) Human parvovirus B19. Historical and clinical review. Rev Inf Dis 10:1005–1011

Turton J, Appleton H, Clewley JP (1990) Similarities in nucleotide sequence between serum and faecal human parvovirus DNA. Epidemiol Infect 105:197–201

Umene K, Nunoue T (1990) The genome type of human parvovirus B19 strains isolated in Japan during 1981 differs from types detected in 1986 to 1987: a correlation between genome type and prevelance. J Gen Virol 71:983–986

Van-Elsacker-Niele AM, Salimans MM, Weiland HT, Vermey-Keers C, Anderson MJ, Versteeg J (1989) Fetal pathology in human parvovirus B19 infection. Br J Obstet Gynaecol 96:768–775

Veijalainen P (1988) Characterization of biological and antigenic properties of raccoon dog and blue fox parvoviruses a monoclonal antibody study. Vet Microbiol 16 (3):219–230

Veijalainen PML, Neuvonen E, Kangas J (1984) Parvovirus infection in blue foxes. In: Proceedings 3rd international scientific congress on fur animal production, Versailles, Paris, France, April, Institut National de la Recherche Agronomique, paper 58

Walker ST, Feilen CP, Love DN (1979) Canine parvovirus infection. Aust Vet Pract 9:151–153

Walker ST, Feilen CP, Sabine M, Love DN, Jones RF (1980) A serological survey of canine parvovirus infection in New South Wales, Australia. Vet Rec 106:324–325

Waller EF (1940) Infectious gastroenteritis raccoons (= Procyon Loter). J Am Vet Med Ass 96:266

Whur P (1985) Parvovirus vaccination. Correspondence. Vet Rec 117 (1):22

Williams MD, Cohen BJ, Beddall AC, Pasi KJ, Mortimer PP, Hill FG (1990). Transmission of human parvovirus B19 by coagulation factor concentrates. Vox Sang 58:177–181

Wilson ND (1980) Origin of canine parvovirus. Vet Rec 106:392

Wong FC, Spearman JG, Smolenski MA, Loewen PC (1985) Equine parvovirus: initial isolation and partial characterisation. Can J Comp Med 49 (1):50–54

Yasuhara H, Matsui O, Hirahara T, Ohgitani T, Tanaka M, Kodama K, Nakai M, Sasaki N (1989). Characterization of a parvovirus isolated from the diarrheic faeces of a pig. Nippon Juigaku Zasshi 51:337–344

Zschokke E (1900) Über Coli-bacillare Infektionen. Schweiz Arch Tierheilk 42:20